EAI/Springer Innovations in Communication and Computing

Series Editor

Imrich Chlamtac, European Alliance for Innovation, Ghent, Belgium

Editor's Note

The impact of information technologies is creating a new world yet not fully understood. The extent and speed of economic, life style and social changes already perceived in everyday life is hard to estimate without understanding the technological driving forces behind it. This series presents contributed volumes featuring the latest research and development in the various information engineering technologies that play a key role in this process.

The range of topics, focusing primarily on communications and computing engineering include, but are not limited to, wireless networks; mobile communication; design and learning; gaming; interaction; e-health and pervasive healthcare; energy management; smart grids; internet of things; cognitive radio networks; computation; cloud computing; ubiquitous connectivity, and in mode general smart living, smart cities, Internet of Things and more. The series publishes a combination of expanded papers selected from hosted and sponsored European Alliance for Innovation (EAI) conferences that present cutting edge, global research as well as provide new perspectives on traditional related engineering fields. This content, complemented with open calls for contribution of book titles and individual chapters, together maintain Springer's and EAI's high standards of academic excellence. The audience for the books consists of researchers, industry professionals, advanced level students as well as practitioners in related fields of activity include information and communication specialists, security experts, economists, urban planners, doctors, and in general representatives in all those walks of life affected ad contributing to the information revolution.

Indexing: This series is indexed in Scopus, Ei Compendex, and zbMATH.

About EAI

EAI is a grassroots member organization initiated through cooperation between businesses, public, private and government organizations to address the global challenges of Europe's future competitiveness and link the European Research community with its counterparts around the globe. EAI reaches out to hundreds of thousands of individual subscribers on all continents and collaborates with an institutional member base including Fortune 500 companies, government organizations, and educational institutions, provide a free research and innovation platform.

Through its open free membership model EAI promotes a new research and innovation culture based on collaboration, connectivity and recognition of excellence by community.

More information about this series at http://www.springer.com/series/15427

Mallikka Rajalingam

Text Segmentation and Recognition for Enhanced Image Spam Detection

An Integrated Approach

Springer

EAI
RESEARCH MEETS INNOVATION

Mallikka Rajalingam
Department of Computer Science & Engineering
Bharathidasan University
Tiruchirappalli, India

ISSN 2522-8595 ISSN 2522-8609 (electronic)
EAI/Springer Innovations in Communication and Computing
ISBN 978-3-030-53046-4 ISBN 978-3-030-53047-1 (eBook)
https://doi.org/10.1007/978-3-030-53047-1

This Springer imprint is published by the registered company Springer Nature Switzerland AG
The registered company address is: Gewerbestrasse 11, 6330 Cham, Switzerland

Preface

This book has proposed an efficient spam detection technique which is a combination of character segmentation, recognition and classification (CSRC) that could detect whether an email (text- and image-based) is a spam mail or not. The present work is presented with a fourfold process. First, the text character is extracted from the image by segmentation process which includes a combination of discrete wavelet transform (DWT) and skew detection. Thus, the image features of a specific shape can be isolated and the regular curves such as circles, lines and ellipses can be detected. Second, the text characters are recognized via text recognition and visual feature extraction approach which relies on contour analysis with improved local binary pattern (LBP). Third, the extracted text features are classified using improvised K-nearest neighbour search (KNN) and support vector machine (SVM) classifiers, and the text data for both classification and regression process are analysed. Fourth, the performance of the proposed method is validated by the measure of metrics such as sensitivity, specificity, precision, recall, F-measure, accuracy, error rate and correct rate.

Contents

Chapter 1
Introduction

1.1 Introduction

With the present advancement in internet, there is an increased utilization of email communication which has become one among the fastest modes of communications. However, an increase in the usage of email communication has led to the increased rate of spam-based issues all over the world. According to Rekha and Negi [Rek, 14] around 90% of emails that arrive at the mailbox of email users are spam emails wherein these emails contain junk information that tends to affect the normal computing utilities of email users. While spam emails are generally based on advertising content, in many cases they also contain malicious code and virus which might harm the users' account [Fir, 10]. With advancements in the technologies to detect email spam emerged, spammers developed the concept of image spamming which tends to complicate the processes of detecting spam in image mails. Though previous researchers attempted to develop novel techniques for the detection of image spam, there is still a gap to develop an efficient image spam detection system which could detect spam in images wherein the scalability of the method should improve despite the type of image spam that is sent. In this regard, the present research in this chapter introduces a brief overview about the research topic, a clear understanding on the concepts of spam detection, the characteristics of spam detection, the challenges in development of image spam detection, the contribution of the research and outline that will be followed in the research.

Email communication is the most prominent way of communicating with others. Global email account raised from 3.3 billion in 2012 to 4.3 billion in 2016 [Rad, 12] with 6% yearly growth rate. In this regard with such an alarming usage of email communication, managing emails against fraudulent activities has become an important task. The unwanted emails which are sent to the users are considered as spam messages. A spam mail is defined as an unsolicited/irrelevant/unwanted mail message received by users [Kam, 10]. Spam emails usually contain commercial or

M. Rajalingam, *Text Segmentation and Recognition for Enhanced Image Spam Detection*, EAI/Springer Innovations in Communication and Computing,
https://doi.org/10.1007/978-3-030-53047-1_1

profitable campaigns of uncertain products, dating services, get-rich-quick schemes and advertising. Spam emailing is also used to spread malicious or virus codes and is intended for fraudulence in financial transaction or phishing. Spamming is considered to regulate losses over the internet especially when they tend to turn malicious for business organizations. Several losses are mostly collateral damages not focusing on a particular network or any organization. Spam emails occupy more network bandwidth during transmission. It also consumes user time in terms of searching. Statistical reports show, as of December 2014, spam messages accounted for 66.41% of email traffic worldwide, and Asia constitutes 54% of the total percentage [Sta, 17]. A recent study by Biggio et al. [Big, 11] reveals the fact that most of the users receive more spam emails than non-spam emails.

Spam is unwanted, unsolicited commercial email and messages sent massively, directly or indirectly, by an unknown sender which clutters inbox and affects email server [Bos, 14; Das, 14]. Emails that the recipient does not like to get are called spam emails. A huge number of same messages is forwarded to many receivers by email. Growing amount of similar spam emails is creating grave issues for internet service providers, internet users and the entire internet backbone network. One of the instances of this may be refusal of service where spammers cause an enormous traffic to an email server, thereby deferring valid messages to reach planned receivers. Spam emails not just squander sources like bandwidth, storage space and computation force, but may include deceitful plans, false proposals, and strategy. Other than that, the time and zeal of email recipients is squandered. Because they have to trace valid emails among the spam and take steps to get rid of the spam, it is an extremely hard job to handle spam and categorize it. Besides, one pattern cannot deal with the issue as fresh spam is continuously coming up and these spams are time and again energetically customized. Therefore, they are not discerned including furthermore obstacle to exact discovery [Rek, 15].

1.2 Characteristics of Image Spam

Though text-based spam emails are detected by most methods of email spam detection, spammers have identified new routes towards sending spam messages through images. Such a form of sending spam messages through images is called as image spamming, and images embedded with spam characteristics are known as spam images or Image spam. Most algorithms find it easy to identify spam in text email. However, the same in image spam emails is a daunting task. A spam image carries a message which is intended to reach client systems and displays the same. One another complexity of spam detection techniques is though they are better methods to detect spam; they may also intend to block ham messages wherein the process is known as false positive [Meh, 08]. The characteristics of image spam are shown in Table 1.1.

However, detection of image spam is a difficult task as the messages or token (characters) is embedded within the images. The token or character embedded in the image needs to be extracted and should be converted (also known as character recognition)

Table 1.1 Characteristics of Image spam

Characteristics	Description
Image spam is text messages with noise	All spam image emails contain text messages which are intended to depict the information shared by the spammer. Most spam images are advertisements and are generally blacklisted (e.g. Cialis pills, drug store, stock tip)
Image spam are distinct from one another	Spammers take utmost care to uniquely design each spam image using sophisticated algorithms so as to ensure that the image spam is unique. Several techniques have been utilized so as to arrange the elements in the image spam email such as noise, background, colours of the fonts and so on. Adding these features transform the image spam complex to be identified [Fum, 06]
I-spam messages use HTML effectively	I-spam utilizes MIME for transporting the attached image data with the HTML formatting and text which is non-suspicious. Such text is different from what is actually present in the email
I-spam messages are different from natural images	The colour space of natural images is smooth and hence is distinct from image spam messages which are generally sharp and clear objects

into ASCII form. Character recognition within an image is indeed a challenging task as it involves image processing as the first process which involves character segmentation to mark the character in the image and the second process known as character recognition which is to convert the marked character into ASCII form. In the final process, ASCII forms are ready to be processed for identifying spam emails. Detecting spam emails especially image spam as shown in Fig. 1.1 is the focus of the present research which is a challenging task when compared with other conventional spam detection techniques.

1.3 Problem Statement

The problem of spam detection has acquired immense attention wherein specific challenges such as text classification or categorization require attention. Though researchers have addressed such challenges in a more generic manner, following are the problems faced:

1. Spammers all over the world tend to create new techniques to spam through images and text.
2. Text embedded in images were subjected to noise such as background pattern, colour, font variations and imperfections in a font size so as to eliminate the chances of being identified as spam by filtering techniques.

Hence, an algorithm to appropriately detect image spam emails should be proposed which became the premise of the present research; however, this requires the combination of one or more algorithms and the development of a system which could appropriately detect image-based spam mail. In this regard, any image-based

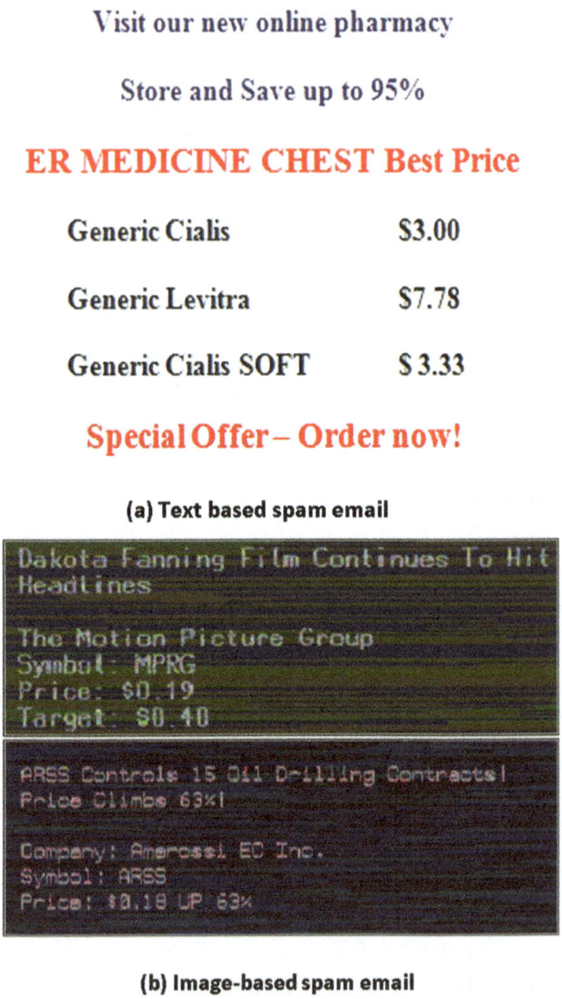

Fig. 1.1 Sample spam email. (**a**) Text-based spam email; (**b**) image-based spam email

spam detection method takes into consideration three major processing steps that could regulate image spam detection. Firstly, character segmentation is the preliminary task performed in the process flow of spam detection. Character segmentation is the process that marks or segmented every character in the image. According to Casey and Lecolinet [Cas, 96], character segmentation is a procedure in which a considered image is decomposed into sub-images possessing individual symbols of the text. Character segmentation which is the first procedure in the proposed system should take into consideration several criteria which are as follows: Source adopted from [Cas, 96].

Steps for Character Segmentation

1. Identify the pattern of characters provided in image spam with the resemblance of symbols in a system.
2. Character pattern matches should be appropriate. For example, both 'cl' and 'd' may look the same in an image.
3. Cursive character patterns should also be identified accurately.

Owing to the text differences including style, size, alignment, less contrast and complex background image, segmentation technique turned into an exigent task which implies the need for an algorithm that could detect line and curve separating each alphabet in the image.

Once each character/object in the image is segmented, the next step is to identify the marked object and change to character (ASCII form). This is known as character recognition. Character recognition is a technique which involves classification of input formation on the basis of requirements of the systems which are imposed during such classification. Character recognition is performed with the context that not always shall the decision taken for recognition is accurate, but character recognition techniques should impart some algorithms that could recognize a character with better accuracy. This is better explained as follows:

'Assume a set M of objects which are segregated into n- different non-intersecting subsets known as characters or object classes. Each character is designated by a character description x which should be compiled as a multi-dimensional vector. Object description should not necessarily be unique and may correspond with other classes of objects' [Nad, 15]. In general, characters are typically monotonic on a fixed background, and hence character recognition in images is potentially far more complicated which includes other possible variations such as changes in background, lighting, texture and font.

Once character segmentation and character recognition are fully operational, the next step will be to combine them as a single image spam detection system. The combined system should enable identification of an image mail as Ham or Spam. The refined extracted characters should be preprocessed for email detection.

1.4 Objectives

The objectives of this research are mentioned as follows:

- To design an efficient character segmentation, recognition and spam detection algorithm for the segmentation and recognition of image spam emails using improvised DWT, Hough transforms along with spatial frequency cross-correlation for automatic segmentation, contour analysis with an improved local binary pattern for text recognition and improvised SVM and KNN classifiers for visual feature extraction.

- To analyse the proposed technique's performance using precision, f-measure, recall and accuracy.
- To evaluate the limitations of the proposed research thereby recommending future researches.

1.5 Motivation

The number of spam messages are increasing in present days that hinder the normal operations of mail users. With the development of new techniques to restrict text-based spam messages, spammers identified new techniques wherein spam images are embedded in images and are sent to email users. Though there is immense literature that attempted to mitigate the issues arising out of image spam, there is still an unaddressed gap which is the inability of algorithms and techniques proposed to identify spam emails from legitimate emails. A need persists to devise a novel technique which could recognize image spam emails which motivated the researcher to identify the various techniques used till date and the development of a novel algorithm-based technique to recognize ham and spam image mails.

1.6 Research Contribution

Nowadays the number of online spamming cases is increasing which is hazardous to safe internet utilization. Spam that is created in excessive amounts is an issue for the reduction of information quality and creates a concern for web users. Spammers utilize image-based email for the collection of private data and perform phishing attacks. There is hence a need to develop a system which could appropriately detect image spam and neglect ham images which is the motivation for the present research. In this context, various techniques were examined from the literature, and the solution to image spamming issue is the combination of various techniques which on the whole could contribute to better image spam identification.

The current work has a major contribution to developing an image-based spam detection system which is a combination of character segmentation, recognition and classification (CSRC) that could detect whether an email (text and image based) is a spam mail or not. In this regard, the present research is presented with three methods. The proposed methods are distinct to each other and present three contributions to the body of knowledge thereby achieving the outlined research objectives. The contributions of this study are as follows:

- This study has the major motive which is to solve the spam detection from the image which can be easily by proposing a novel image spam filtering technique that is scalable and adaptable. The framed detection approach made extraction of embedded text along with colour, texture, shapes which are utilized to estimate similarity with the query image. The feature extractions are utilized to train the classifier that classifies the online message as spam or authorized.

- We propose a novel unified-step framework in image spam detection based on the combination of robust and improvised DWT, Hough transforms along with spatial frequency cross-correlation for automatic segmentation, while contour analysis with an improved local binary pattern for text recognition. Visual feature extraction using improvised SVM and KNN classifiers. Thus, the present research proposed a spontaneous, constant, rapid response automatic segmentation, feature extraction and classification to detect spam from the images and the text. The proposed method was compared with other traditional methods.
- A novel algorithm DWT with skew detection for character segmentation was proposed. Character segmentation from images are done using DWT, which includes morphological dilation operators and the logical AND operators to remove the non-text regions, and Hough transforms along with spatial frequency cross-correlation. Further, to reduce the size of images, skew detection specifically applying a fusion of Hough transform with spatial frequency cross-correlation was proposed. Previously skew detection algorithms such as Hough transforms, clustering, projection profiles, wavelet decompositions, morphology, moments, space parallelograms and Fourier analysis work on the assumption that images are black and white and enhanced for documents among which text is prominent and aligned in the form of parallel straight lines. However, previous algorithms could not make an exact solution in case of its usage in suitable documents. For skew detection, specifically, Hough transform with spatial frequency cross-correlation was proposed. The fusion-based proposed method considers polygons. Image's structure or texture and threshold for separating it into polygons or connected areas.
- The research proposed contour analysis with an improved local binary pattern for text recognition and visual feature extraction. To acquire the image's smooth contours, double filter bank, Laplacian pyramid (LP), directional filter bank (DFB) provide better multiscale decomposition and remove the low frequency. LBP considers the effects of central pixels, and presents complete structure patterns to enhance the discriminative ability.
- The extracted features are classified using SVM with a KNN classifier. KNN was used to extract features by predicting the nearest neighbour SVM and analyse the data for classification and regression.
- The proposed methods have both training and testing phase.

1.7 Research Scope

The goal of this research is to improve the accuracy of email spam detection. More precisely, the present research tends to assess the different methods that are capable of identifying individual text and image-based emails; however, image-based spam detection is the main focus of the research. The project hence limits its scope towards identifying image-based spam emails and does not intend to identify the entity that actually spreads spam messages. Email legitimacy is determined by the proposed approach. Furthermore, the proposed approach is a new contribution to secure email usage as detection accuracy of proposed technique outperformed the existing approaches.

1.8 Novelty and Significance

The present research has its novelty towards manipulating several techniques of character segmentation and recognition wherein spam images are recognized using shape-based feature extraction methods. Such combination of techniques such as DWT and Hough transforms, and Template matching and Contour analysis is a relatively new method in the field of research wherein the proposed model is also hypothesized to bring better results in terms of accuracy of spam detection. This method is also significant towards bringing insights for future researchers to conduct research. However, for the improvisation of the performance of the segmentation and recognition processes, additional methods are used such as spatial frequency cross-correlation, improved local binary pattern and so on.

1.9 Outline of the Chapters

The book is organized into eight chapters with appendices.

Chapter 1 outlines an introduction to the text and image-based email classification, followed by the motivations and problem statement, research objectives, research scope, novelty and significance of the research.

Chapter 2 is the literature review to identify the strengths and limitations of the current text and image-based email classification approaches wherein previous researchers are assessed and explored. It first elaborates on the concepts and definitions pertaining to the present research topic and describes the overview of spam detection, character segmentation and character recognition methods. Furthermore, the chapter provides a detailed description of image-based email detection techniques. The limitations of previous researches examined are detailly explained in the research gap and the summary of the chapter.

Chapter 3 covers the information on the data sets used, and the steps involved in preprocessing performed in the present research.

Chapter 4 presents in detail an enhanced character segmentation algorithm that improves the detection efficiency of image-based emails using the hybrid approach of DWT and Hough transform methods with pixel count analysis technique; in addition, the researcher used Spatial frequency cross-correlation to improve the processes of segmentation wherein text is segmented from the image-based email efficiently. The component of the algorithm and their functions are discussed.

The experimental results of this algorithm are also presented and compared with related methods of the literature in terms of segmentation accuracy wherein the performance of the algorithm used is assessed.

Chapter 5 presents in detail the processes involved in character recognition. The segmented characters are corrected using skew detection and correction. The combined approach of template matching and contour analysis is used to recognize the character wherein error corrections and improved local binary pattern will be applied. The components of this algorithm and their functions are discussed. In addition, the experimental outcomes of the framed technique are presented and collate with related methods in terms of recognition accuracy which is a means to examine the accuracy of the proposed algorithm.

Chapter 6 presents in detail a detection algorithm for image-based ham/spam emails using classification/feature extraction using SVM and KNN classifier. The structure and texture of an image will be examined, and the detection technique encompasses optimisation, nearest neighbour search, handling inconsistent constraints and error corrections. The proposed technique's performance is also assessed.

Chapter 7 discusses the entire approach with the discussion of the different algorithms used followed by testing the entire system based on parameters such as False Positive (FP), False Negative (FN), True Positive (TP), True Negative (TN), Recall and Precision which are used to evaluate the performance of the proposed work.

Chapter 8 furthermore concludes the investigation and suggests recommendations for the upcoming task with esteem to this research.

'Appendix' section covers snippets of code used in the image-based ham/spam detection approach.

References

[Rek, 14] Rekha, & Negi, S. (2014). A review on different spam detection approaches. *International Journal of Engineering Trends and Technology, 11*(6), 315. Retrieved from http://www.ijettjournal.org/volume-11/number-6/IJETT-V11P260.pdf.

[Fir, 10] Firte, L., Lemnaru, C., & Potolea, R. (2010). Spam detection filter using KNN algorithm and resampling. In: *Proceedings of the 2010 IEEE 6th International Conference on Intelligent Computer Communication and Processing*. [Online]. August 2010, IEEE. Retrieved from http://ieeexplore.ieee.org/document/5606466/.

[Rad, 12] Radicati, S., & Hoang, Q. (2012). *Email statistics report*. [Online]. PALO ALTO. Retrieved from http://www.radicati.com/wp/wp-content/uploads/2012/04/Email-Statistics-Report-2012-2016-Executive-Summary.pdf.

[Kam, 10] Kamboj, R. (2010). *A rule based approach for spam detection*. Patiala: Thapar University.

[Sta, 17] Statista. (2017). *Global spam volume as percentage of total e-mail traffic from January 2014 to September 2016, by month*. [Online]. 2017. The Statistics Portal. Retrieved January 3, 2017, from http://www.statista.com/statistics/420391/spam-email-traffic-share/.

[Big, 11] Biggio, B., Fumera, G., Pillai, I., & Roli, F. (2011). A survey and experimental evaluation of image spam filtering techniques. *Pattern Recognition Letters, 32*(10), 1436–1446. Retrieved from http://linkinghub.elsevier.com/retrieve/pii/S0167865511000936.

[Bos, 14] Bosworth, S., Kabay, M. E., & Whyne, E. (2014). *Computer security handbook, set* (6th ed.). New York: Wiley.

[Das, 14] Das, M., & Prasad, V. (2014). Analysis of an image spam in email based on content analysis. *International Journal on Natural Language Computing, 3*(3), 129–140. Retrieved from http://www.airccse.org/journal/ijnlc/papers/3314ijnlc13.pdf.

[Rek, 15] Rekha, & Negi, S. (2015). A review on different glaucoma detection. *International Journal of Engineering Trends and Technology., 11*(6), 2–7.

[Meh, 08] Mehta, B., Nangia, S., Gupta, M., & Nejdl, W. (2008). Detecting image spam using visual features and near duplicate detection. In *Proceeding of the 17th international conference on World Wide Web—WWW '08* (pp. 497–506). New York, NY, USA: ACM Press. Retrieved from http://portal.acm.org/citation.cfm?doid=1367497.1367565.

[Fum, 06] Fumera, G., Pillai, I., & Roli, F. (2006). Spam filtering based on the analysis of text information embedded into images. *Journal of Machine Learning Research, 7*(1), 2699–2720. Retrieved from http://www.jmlr.org/papers/volume7/fumera06a/fumera06a.pdf.

[Cas, 96] Casey, R., & Lecolinet, E. (1996). A survey of methods and strategies in character segmentation. *IEEE Transactions on Pattern Analysis and Machine Intelligence, 18*(7), 1–31. Retrieved from http://perso.telecom-paristech.fr/~elc/papers/pami96.pdf.

[Nad, 15] Nadeem, D., & Rizvi, S. (2015). *Character recognition using template matching*. New Delhi: Jamia Millia Islamia. Retrieved from https://pdfs.semanticscholar.org/c1b5/dcd918da02f72a9579ed5eeeab111da3c7cb.pdf.

Chapter 2
Review of Literature

This chapter encompasses the overall review of spam emails and the variety of existing email classification techniques used for spam detection with intense analysis of their strengths and weaknesses. Furthermore, the chapter elucidates the concept of text-based email classification with various machine learning approaches including Naïve Bayes, Decision Tree and SVM (Support Vector Machine). The chapter also presents a detailed description of image segmentation, character recognition and image-based email detection which is deliberated as the foundation of the present research.

2.1 Character Segmentation

Segmentation technique divided the digital image into various multiple segments. The major motive of this technique is to demonstrate the image in a simple manner and for the easier analysing purpose. Segmenting cursive characters from the image is a difficult task, and segmenting low-resolution characters is also challenging in document image processing. Detecting empty space from the document image is a problematic responsibility in character segmentation process.

The study of machine learning is a subfield from artificial intelligence, with an intention to make technologies capable of absorbing like that of a human brain. Knowledge through machine learning means to observe, understand and signify information about the statistical occurrence. Unsupervised learning algorithms try to find out the unseen orderliness (clusters) or identify the abnormalities in data such as spam messages or network interference. In the filtration of emails there may be some sack of words or the subject-line investigation. There are two important features in email classification which are typically separated into numerous subtasks. Firstly, group of data and its demonstration are regularly problematic in particular

M. Rajalingam, *Text Segmentation and Recognition for Enhanced Image Spam Detection*, EAI/Springer Innovations in Communication and Computing,
https://doi.org/10.1007/978-3-030-53047-1_2

(i.e. email messages). Secondly, email feature selection and future deduction challenge to decrease the features quantity for durable task steps. Authentic mapping within the training and testing set has been identified by email classification phase. Machine learning techniques used to serve the aforementioned tasks are elaborated in the following section. Character segmentation is categorized into two subsections: classifier-based and non-classifier-based techniques.

2.1.1 Classifier-Based Approach

2.1.1.1 Naive Bayes Classifier

The first Naïve Bayes classifier for spam recognition is proposed in 1998. Bayesian classifier operates on proceedings which are dependent and possibility of event which may occur in the future or which can be identified after the earlier happening of the similar occasion [Alm, 11]. This procedure sorts spam emails by analysing the words in the mail as a central rule and also checks for the words that have been frequently occurring in the spam and in the ham. If there are repetitive words found in the mail, then the received email is declared as spam. Naïve Bayes classifier method has been developed as a widespread technique for email filtration, and Bayesian filter is qualified to work successfully. Each word has certain possibilities that occur in the ham or spam email in the database. Doubt in the chances of the whole count of words exceeds a definite border; the filter places the email to any group (ham/spam). There are just two groups of emails. It could be either a spam or ham. In effect, all the spam filters grounded on stats employ Bayesian likelihood computation to include definite token's info to a universal score [Awa, 11a]. On the basis of the universal score, conclusions are drawn. The info is generally interesting for a token T because of its spamming or rating of spam (score) that is computed as shown below:

$$S[T] = \frac{C_{\text{spam}}(T)}{C_{\text{spam}}(T) + C_{\text{Ham}}(T)} \tag{2.1}$$

where $C_{\text{spam}}(T)$ and $C_{\text{Ham}}(T)$ are the count of spam or ham messages including token T, respectively. To estimate the chance for a message M with tokens $\{T_1 \ldots T_N\}$, individual desires to connect the individual token's spamming to determine the whole message spamming [Awa, 11a]. A humble method to produce categorizations is to estimate the item of individual token's spamming and to match it with the item of individual token's hamminess that is furnished below:

$$\left(H[M] = \prod_{I=1}^{N} \left(1 - S[T_I]\right) \right) \tag{2.2}$$

An email communication is indicated as spam if the comprehensive spamming items $S[M]$ is higher than the hamminess items $H[M]$. The overhead interpretation is employed in the algorithm mentioned below [Awa, 11b].

Stage 1: Training

Examine for every email into its essential tokens.
Create a probability for every token W

$$S[W] = C_{\text{spam}}(W) \Big/ \left(C_{\text{ham}}(W) + C_{\text{spam}}(W)\right) \tag{2.3}$$

Store spamminess values to a database.

Stage 2: Filtering

For each communication or message M.
While (M not end) do
Scan message for the following token T_i
Request the database for spamminess $S(T_i)$
Compute collected communication or message probabilities
$S[M]$ and $H[M]$.
Compute the overall communication or message filtering sign by:
$I[M] = f(S[M], H[M])$.
f is a filter-dependent function,
such as $I[M] = (1 + S[M] - H[M])/2$.
If $I[M] >$ threshold
msg is noticeable as spam
else
msg is noticeable as ham

Researcher [Vij, 18] used Naïve-based classifier with three-layer framework for detecting bulk spam email. They experimented with real-time data set for the detection of legitimate and spam email. To improve the accuracy, feature extraction was used to extract the features based on bucket classification. Self acknowledgeable Internet Mail System was implemented to know the status of the sender mail.

2.1.1.2 *K*-Nearest Neighbour Classifier

The *k*-nearest neighbour (*K*-NN) classifier is an example grounded one that employs the instructing records for the evaluation than an obvious group exhibition that includes the classification summaries employed by earlier classifiers; however, there is no real preparation phase. Whenever there is a necessity to categorize a new record, the *k* most similar records (neighbours) are created to be verified, if the large section of them are assigned to a certain group, the new record is assigned to this group. Besides, finding out the nearest neighbours can quicken the employment of conventional indexing procedures. The categorization of spam or ham messages is determined with the category of the messages that are so near to it. The assessment among the vectors is a real method [Ger, 17]. This is the notion of the *k*-nearest neighbour algorithm:

Stage 1: Training

Keep the training email or communication.

Stage 2: Filtering

Provide a communication or message *x*, command its *k*-nearest neighbours betwixt the emails in the preparation group. If there is further unsolicited email or spam preparation group or if there are further spam betwixt these neighbours, categorize provided email or message as spam. Or else, categorize it as ham.

Indexing procedure decreases the duration of comparability with the intricacy of $O(m)$, where m refers to model dimension. The procedure is in addition mentioned as memory-grounded classifier as all the preparation instances are saved in the memory [Pat, 13b; Kho, 07]. One issue with the submitted algorithm is that there is no such criterion which could perhaps decrease the count of fake positives. However, the issue may simply be dealt with by altering the categorization principle to the l/k-rule mentioned below: If l or further messages among the *k*-nearest neighbours of x are spam, categorize x as spam, or else, categorize it as valid emails. The *k*-nearest neighbour principle has detected broad usage in common categorization jobs. In addition, it is one of the scarce globally persistent categorization principles.

2.1.2 Artificial Neural Networks Classifier

An artificial neural network (ANN), alias 'Neural Network' (NN), is a computational grounded simulation of biological neural network methods that functions on the rule of studying by instance [Meh, 17]. NN is a flexible arrangement that comprises an interrelated gathering of artificial neurons in order to alter its framework grounded on info that runs by means of the network throughout the studying stage. Although there are several types of neural networks, the traditional kinds are perceptron and multilayer perceptron.

The concept of the perceptron is to find a linear use of the feature vector $f(x) = wTx + b$ in a manner that $f(x) > 0$ for vectors of one category [Mar, 09] and $f(x) < 0$ for vectors of other category. Here, $w = (w_1, w_2, \dots w_m)$ is the vector of coefficients of the formula, and b is bias. The algorithm indicates the categories by numbers +1 and −1, declares that hunt for a decision function $d(x) = \text{sign}\ (wTx + b)$. The perceptron studying is carried out with an iterative algorithm begins with arbitrarily selected criteria (w_0, b_0) of the conclusion and updates them iteratively. On the nth iteration of the algorithm, a training sample (x,c) is selected in a manner that the present decision function does not categorize it properly (i.e. $\text{sign}(w_n x + b_n) \neq c$). The criteria ($w_n$,$b_n$) are then updated employing the principle:

$$w_n + 1 = w_n + c_x \tag{2.4}$$

$$b_n + 1 = b_n + c \tag{2.5}$$

The algorithm ceases when a decision function is detected that rightly categorizes all the training samples. The above explanation is employed in the algorithm given below [Car, 06]:

2.1.2.1 Stage 1: Training

Initialize w and b (to random value or to 0).
Determine a training example (x,c) for which sign ($w^T x + b$)
If there is no such example, then training is finished
Save the final w and halt
Or else go to next step
Update (w,b): $w := w + c_x$, $b := b + c$ Go to earlier step.

2.1.2.2 Stage 2: Filtering

Provided a message x, find its category as sign ($w^T x + b$)

2.1.3 *Support Vector Machines Classifier*

Support Vector Machines are applied with the idea of conclusion planes that describe conclusion borders [Tia, 12]. A conclusion plane divides group of objects as categories possessing various memberships at which point the SVM modelling algorithm ascertains a maximum hyper plane with the highest edge to divide two categories that needs unravelling the optimization problem mentioned below:

Maximize

$$\sum_{i=1}^{n} \alpha_i - \frac{1}{2} \sum_{i,j=1}^{n} \alpha_i \alpha_j y_i y_j K(x_i, x_j) \tag{2.6}$$

Subject to

$$\sum_{i=1}^{n} \alpha_i y_i = 0$$

$$\text{where } 0 \leq \alpha_i \leq b,\ i = 1,2,\ldots.n$$

where α_i is the weight of training sample x_1. If $\alpha_i > 0$, x_1 is known as a support vector, b is a principle criterion employed to compromise the training precision and the sample intricacy in order to attain the supreme generalization ability. The kernel function K gauges the conformity betwixt two models. A famous radial basis function (RBF) kernel function,

$$K\left(x_i, x_j\right) = \exp\left(-\gamma x_i - x_j^{2}\right), \gamma > 0$$

After the weights are determined [Say, 11], a test sample x is classified by

$$y = sign\left(\sum_{i=1}^{n} \alpha_i y_i K\left(x_i, x_j\right)\right),$$
$$\text{Sign}\left(a\right) = \begin{cases} +1, & \text{if } a > 0 \\ -1, & \text{otherwise} \end{cases} \tag{2.7}$$

A cross-verification procedure is carried out to ascertain the values of $<\gamma, b>$ on the training data set. Cross-verification calculates the generalization ability on fresh models which are not in the training data set. A k-fold cross-verification unsystematically divides the training data set into k roughly same-sized subsets, drops one subset, constructs a classifier on the balance models in order to assess the categorization execution on the new subset [And, 17]. This procedure is iterated k times for every subset to attain the cross-verification execution over the entire training data set. If the training data set is huge, a diminutive subset can be employed for cross-verification to reduce calculating charges. The algorithm mentioned below can be employed in the categorization procedure.

Input: Sample x to classify.
Training set $T = \{(x_1,y_1), (x_2,y_2), \ldots (x_n,y_n)\}$;
Number of nearest neighbours k.
Output: Decision $yp \in \{-1,1\}$.
Find k sample (x_i,y_i) with minimal values of $K(x_i,x_i) - 2 * K(x_i,x)$.
Train an SVM model on the k-selected samples.
Classify x using this model, get the result yp.
Return yp.

2.1.4 Decision Tree

A decision tree is a prognostic sample that widens a tree of decision and their likely effects, containing possibilities of event results and source rates. The result of the decision tree can be separate or as in case of regression trees conjunction of elements results in the categorizations at diverse stages [Sar, 12]. Prevalent decision tree studying procedures are C4.5, ID3 and J48.

The decision tree produced by C4.5 can be employed for diverse categorization issues. The algorithm selects a quality at every node of the tree which can furthermore divide the models into subsets. Every leaf node depicts a categorization or conclusion. Certain premises direct this algorithm, like the ones listed below [Chr, 10]:

- If all instances are of the identical category, then the tree is a leaf and hence the leaf is given back with the marked category.
- Compute the possible info for every quality (grounded on the chances of every instance possessing a specific value for the quality).
- Compute info gain for every quality (grounded on the chances of every instance with a specific value for the quality being of a specific category).
- Relying on the present choosing parameter, ascertain the best quality to branch on.
- J48 is an open resource execution of C4.5. Decision tree is constructed by examining data nodes that are employed to assess the importance of present elements.

J48 constructs decision trees from a group of training stats employing the idea of info chaos. J48 verifies the standardized info gain that outcomes from selecting a quality for dividing the stats. It employs the reality that every quality of info can be employed to make a conclusion by dividing the stats into smaller subsets. J48 classifier recursively categorizes till each leaf is clean. It means that the stat has been classified as near to ideal as likely [Mah, 13]. Employing the idea of info entropy, J48 constructs decision trees from a group of training stats in the identical method as ID3. The training stats is a set ($S = s_1, s_2, \ldots$) of already categorized models. Every model ($s_i = x_1, x_2, \ldots$) is a vector, where $x_1, x_2, \ldots$ depict qualities or elements of the model. The training stats is increased with a vector ($C = c_1, c_2, \ldots$), where $c_1, c_2, \ldots$ depict the category to which every model belongs. At every node of the tree, J48 selects a quality of the stats which most efficiently divides its group of models into subsets adorned in one category or the other. Its parameter is the standardized info gain (disparity in entropy) that ensues from selecting a quality for dividing the info. The quality with the most standardized info gain is selected to make the conclusion. The J48 algorithm then reoccurs on the smaller sublists [Kum, 17b].

This algorithm has some base instances:

- All the models in the list pertain to the identical category. When this takes place, it plainly produces a leaf node for the decision tree saying to select that category.
- None of the elements gives any info gain. In this instance, J48 produces a decision node higher up the tree employing the anticipated value of the category.
- Case of formerly unseen category confronted. Moreover, J48 produces a decision node higher up the tree employing the anticipated value.

2.1.5 *Non-Classifier-Based Approach*

2.1.5.1 Discrete Wavelet Transform

Character partition algorithm in real-time DSP grounded licence plate employing 2D Haar Wavelet Transform could be employed [Wri, 17]. Improved image borders and enhanced LP area detection for its aptness in real-time application are the components of the algorithm. The Haar WT discerns three kinds of edges employing one filter, whereas conventional procedures like Sobel would need above one mask for the undertaking. DWT is a particular instance of sub-bank filtering and computation carried out employing filter bank. The sign is transferred through high-pass and low-pass filters simultaneously to create filtered yield. The procedure of LP detection is edge detection within LP area by means of greyscale differences to verify the edges. DWT is a specialized case of sub-band filtering and calculation done using filter bank. The signal is passed through high-pass and low-pass filters concurrently to generate filtered output. The LP detection technique is the detection of an edge in LP section through greyscale variation, and the Haar edges are compared with greyscale variations to validate the edges. If edges are matched, then a rectangle of connecting edges is drawn. Histogram analysis verifies the character extraction and computes bounding box. The experimental results showed an improvement in 2D Haar WT of character segmentation. Results proved that the method could identify maximum edges in the image, less noise and increased character segmentation ratio. The challenging factor of character segmentation in licence plate is due to raindrops, number plate broken due to accidents or uneven luminance.

Discrete Wavelet Transform and Gradient method extracted text from images. The input image is preprocessed, and the Daubechies DWT is applied that attains edges and texture in three different types [Sya, 14]. Compared with Haar wavelet, Daubechies wavelet contains higher frequency coefficient spectrum. The signal has been decomposed into LL, HL, LH, HH segments of frequency domains. In high contrast of text region, Gradient difference technique was applied to show the difference from non-text regions. By using Otsu thresholding, non-textual information will be removed. The drawback of the proposed method is lesser pixel value compared with a global threshold value which is observed as noise and made the removal of the text region. However, elimination of false positive remains as a challenging task.

2.1.5.2 Hough Transform

Text segmentation in the document image is based on Hough Transform techniques [Sah, 10]. Image acquisition for document image recognition is digitized through the scanner by manual process. The image is preprocessed to convert colour images to greyscale image. Otsu's method is applied to binarize the image and edges are detected. The Hough transform is implemented for extraction of line and word as a set of connected words and stored as bmp file for performance analysis [Gur, 13].

Generalized Hough transforms (GHT) has been applied for Arabic printed document segmentation [Aye, 17]. The voting process gives the Hough transform forcefulness of missing edge points. Segmenting a character by recognition techniques, an indexed dictionary was created for character recognition. Dynamic sliding window technique is used to recognize cursive Arabic characters. The method is grounded on identifying starting and finishing characters of the sub-words, then middle characters are detected. For every last character saved in the dictionary, the similar method is replicated from left restriction of the starting character to recognize the character in the centre. GHT can be employed in OCR not just to identify characters but in addition to search this particular quality for the Arabic cursive character without renovating in the partition phase [Isl, 16]. For experimentation, Arabic printed characters of different font, and different sizes were used wherein 93% of recognition accuracy was achieved. Ali et al. [Ali, 15] proposed document processing concept using optical character recognition system. This concept works like storing the document in computer storage, then reading the content and finally searching the content. For languages other than English to process the information, they used a software called character recognition system.

2.1.5.3 Integrated Approach

A combined method of Licence plate detection is suggested by Panchal et al. [Pan, 16] employing Harris Corner and character partition from a picture. As the result of open structure, an Automatic Licence Plate Recognition (ALPR) has turned out being a crucial investigation focal point. Many arrangements were presented for licence plate recognition, and each procedure had its own specific aims of concern and restrictions. The important measure in ALPR arrangement is the elaborate constraint of number plate, partition, identification. Harris corner algorithm finishes being energetic in altering movement and brightened lightning circumstances. The accuracy of licence plate limitation is nurtured forward to the partition stage. The partition is carried out by a procedure of linked element study united with pixel count, aspect proportion and height of characters.

The good image and challenged image are taken for experimentation with the outcome of the success rate of segmentation accuracy obtained at 93.84%.

2.1.5.4 Projection Profile-Based Technique

Projection profile-grounded method is a procedure for text partition employed right away in run-length contracted, printed English text documents [Jav, 13]. Line partition is carried out employing the projection profile method. Furthermore, partition into words and characters is achieved by tracking the white runs by the foundation area of the text line. Throughout the procedure, a run-grounded area developing method is used in the special vicinity of the white runs to track the perpendicular gap betwixt the characters. After finding out the character gaps in the whole text line, the understanding of word gap and character gap is carried out by calculating the mean character gap. Consequently grounded on the spatial place of the detected words and characters, their respective contracted portions are taken out. For experimentation, the procedure was tried with 1083 contracted text lines, and F-measure of 97.93% and 92.86%, respectively, for word and character partition are acquired.

A character segmentation procedure employing projection profile-grounded method was originated initially by Rodrigues et al. [Rod, 01]. Primary view decision tree algorithm for cursive script identification grounded on the usage of histogram as a projection profile method was originated. A postal code picture info was scanned and changed into a two-dimensional matrix depiction to be employed with a group of algorithms to give complete scope partition. The problems were related with quality and image handlings such as noise, distortion, variation in style, the shift of the character, size of the character, rotation, variation in thickness and variation in texture. For experimentation, 200-dpi pictures were employed with a total of 4320 digits, presuming 8 by strap at which point the executed algorithm took out 3788 ways properly.

For experimentation, 200-dpi pictures were employed with a total of 4320 digits, presuming 8 by strap at which point the executed algorithm took out 3788 ways properly. A mixture method of text partition employing edge and texture element info was suggested by Patel and Tiwari [Pat, 13a]. The texture elements like homogeneity, difference and vitality for texts are dissimilar from non-text. The texture elements are employed to discern the text area from picture. The edge-grounded textures possess several needed elements. The grade magnitudes generally possess higher values in the edge of the characters, even when the text is embedded in images.

Step 1: Change of colour picture to greyscale of picture employing,

$Y = 0.299$ * Red +0.587 * Green +0.119 * Blue.

Step 2: Edge detection is carried out by 3*3 Sobel operator.

Step 3: A threshold is employed for eradication of feeble edges.

Step 4: The edge picture is separated into non-overlapping blocks of m*m pixels.

Step 5: Compute the mean magnitude per pixel and mean grade magnitude per pixel.

Step 6: Separate the filtered grey picture into m*m non-overlapping slabs.

Here, high-pass filter is employed to quash setting.

Step 7: Estimate the element homogeneity and contrast at 00, 450, 900, 1350 directions for every slab of first picture employing grey level co-happening network.

Step 8: Compute the mean of homogeneity and contrast for every slab.

Step 9: Filter the text slabs employing edge-grounded element and texture elements.

Step 10: Combine the acquired text slabs.

The character identification method is further separated into two wide groups: methods grounded on OCR devices, low-level picture elements and text extraction are debated in the upcoming division.

2.1.5.5 DWT and Hough Transform

Analysts [Raj, 16] submitted a hybrid character partition method integrating discrete wavelet transform (DWT) and Hough transform to take out character from pictures. For training and testing, Ling-Spam Corpus database was employed. Primarily, pictures in colour are transformed into greyscale. Employing Otsu's procedure, the greyscale is changed into binary picture. After that, all linked elements that are below 15 pixels are eliminated from the binary picture. Following binarization, the lines and characters are partitioned by employing the advanced mixture method. The advanced sample was tried for exactness, True Positive, False Positive, True Negative, False Negative, accuracy, recollect, F-measure, and was established to be 100%, 0.99, 0.18, 0.81 and 0.008, respectively. Nevertheless, the dimension of training and testing data set was diminutive.

Study by Karanje and Dagade [Kar, 14] have studied the text detection and extraction from various kinds of pictures such as scene picture, born digital picture, and document picture. The study examines various approaches of detection like edge, colour, edge and colour, texture, corner, mixture procedures.

2.1.5.6 Other Studies

Saha et al. [Sah, 10] have stated a work on line and word partition from digital picture employing Hough transform grounded method. The task employed normal data set published on the site by CMATER, Jadavpur University. The record pictures comprise multi-script printed and handwritten text lines with diversity in script and line spacing in single document picture. The method functions rather adequately when used on mobile camera caught business card pictures with low resolution. The utility of the method is examined by using it in a mercantile undertaking for regionalization of licence plate of vehicles from monitoring camera pictures by the procedure of partition itself. The precision of the method for word partition, as examined experimentally, is 85.7% for document pictures, 94.6% for business card images and 88% for monitoring camera pictures. Nevertheless, the algorithm collapses to partition correctly in case of quite intimately spaced lines. Certain method must be employed as a post-processing measure to segregate touching text lines in the Hough picture. In the procedure of partition, the binarization may also do a critical part. A fine binarization method may eradicate few of the document picture partition issues.

2.2 Character Recognition

Character recognition is the procedure of detecting and identifying fonts or characters from input image and changing them into ASCII or other comparable machine-editable form [Ali, 15]. Character recognition is the procedure of changing images of typewritten, printed text or handwritten text into a system comprehended by machines for the objective of editing, decrease in storing volume and indexing/exploring [Kap, 11]. Identification of character is hard when document employed is of bad quality. Issues take place in identifying a character when the font size is small. Character recognition is a difficult job when various kinds of fonts are employed. In the preprocessing stage, the characters are identified by skew detection and skew rectification.

2.2.1 *Pre-processing*

Al-Shatnawi [Sha, 14] used an efficient skew detection and rectification procedure for Arabic handwritten text line grounded on sub-words bounding. It is grounded on three stages containing preprocessing, skew detection and skew rectification stages. The presented approach approximates a text line built on computing the important point for its sub-words bounding. Then, the text-line elements on the evaluated baseline are arranged. The presented approach was accomplished in 3960 text-line handwritten pictures that were written by 40 writers. With regard to efficiency, it was debated with the parallel projection approach. The recommended approach acquired a precision ratio of 96.15% and takes 6.7 s as mean time. In addition, it can spontaneously detect text baselines of documents with any direction.

Stahlberg and Vogel [Sta, 15] suggested a common and strong procedure that can get by with a broad kind of document kinds and writing arrangements. It employed derivatives in the Hough space to recognize paths with quick variations in their projection profiles. This rule is helpful to recognize the parallel and perpendicular orientation concerning the document. Experimented with recommended procedure on the DISEC '13 data set for document skew detection, the effects are comparable to the best arrangements in the literature. The disadvantages of the recommended arrangement are as follows: when used in big non-text regions with no perpendicular or parallel lines, loud projection profiles are created; italic font types bring about leaning in the perpendicular calculation. By employing discrete Fourier algorithm, skew detection and correction approach are improved. The Fourier change and angle of elevation hypothesis to discern the skewness angle first and then skew rectification algorithm was suggested. Since Fourier change will quicken the algorithm, speed will no more be a problem for the suggested algorithm.

In addition, angle of elevation hypothesis has the capability to discern angle in an effective way. By acquiring a model of 120 various skewed pictures, the suggested algorithm was planned and executed in Matrix Lab. These images contain records possessing various languages like English, Hindi, Punjabi and Pictures. In this method, the picture is de-skewed following rectification employing angel of elevation hypothesis. Nevertheless, generally each method has certain restrictions, like few of them give us tempo but are appropriate just for small text, few give us precise outcomes but are sluggish in tempo. Korchagin [Kor, 12] experimented a method for the mechanical time skew detection and rectification for multi-resource audio-visual info, documented by various cameras/recorders throughout the very occurrence. On the basis of ASR-connected elements, all documented info are effectively tried for possible time skew issue and rectified. The crux of the algorithm is grounded on conceptual time-frequency study with an accuracy of 10 ms. The experimental outcomes proved right time skew detection and eradication in 100% of instances for a real-time data set of 32 split meetings and outdoes the function of quick cross relationship whereas maintaining lower arrangement needs. The disadvantage of lossless rectification is that wrappers are usually compatible just with restricted group of multi-resource study software.

Tanase et al. [Tan, 13] have conducted study on document picture skew detection and rectification. The issue was of crucial significance in the mechanical content alteration arrangement territory, forming libraries digitalization undertakings feasible. The analogy betwixt the primary kinds of skew detection algorithms submits the benefits and drawbacks, along with suggested enhancements. The disadvantage of Hough Transform method is intricacy, and mistake takes place if input page includes images.

2.2.2 OCR-Based Character Recognition

OCR-grounded methods take out and study the text set into fastened pictures. While certain execution of these methods can be seen in mercantile spam filters [Sam, 06] and in open resource ones such as Spam Assassin, they have been probed just in the earlier work [Fum, 06]. Present methods are grounded on similar methods that are employed in spam filters to study the body text of an email: keyword detection and text classification. Keyword detection is a plain procedure to proof the spam mail by verifying the happening of normal keywords that is seen in spam emails. This needs a regular update of the keyword list, and can be readily dodged by tricks like misspelling general 'spammy' words. When used in text taken out from OCR devices, this method shows the similar disadvantages mentioned above, and its capabilities can be sabotaged in addition by OCR mistakes. An execution of this method is the OCR plug-in of Spam Assassin. Its default keyword list is brief, and can be tailor-made by end users. The OCR plug-in has Boolean output that is fixed to true if at least one of the keywords is found out in the picture. The Boolean values are numerically coded as 3 (True) and 0 (False), to be united with the yield of the other Spam Assassin's units.

OCR mistakes are made up for by blurred matching betwixt keywords that decreases the result of misspelled 'spammy' words. This method is utilized by one more Spam Assassin plug-in, Fuzzy OCR. Its result is an actual number that rises as the number of keywords seen in the text taken out by the OCR augments. The basic reason is that the more the keywords noticed, the higher the probability that the reasoned picture implants a spam message.

Text classification probes whether the similar text classification methods used in email's body text can be useful in addition to examine the text taken out by OCR [Fum, 06]. Text classifiers prepared on text approaching from email's body are taken into account and tried on text approaching from both the email's body and enclosed pictures (if any). This approach permits the enhancement of picture spam detection rate by disregarding OCR mistakes. The reflection involves that the picture spam detection rate is enhanced when the text taken out by OCR is processed by a different text classifier. Because the text employed for preparing and testing is influenced by similar types of OCR mistakes [Itt, 95].

Assassin plug-in, by providing text to text classifier and the output attained from this classifier is a real number in [0; 1] inferred as the possibility of image embedded spam message which is multiplied with the 4.5 default weight needed to merge with another spam assassin module's outputs.

2.2.3 Low-Level Image Features

Fumera et al. [Fum, 06] developed a technique for anti-spam filtering process by utilizing text data inserted within images that are sent as supplements. This technique utilizes the text categorization for examining the text data obtained from attached images using OCR tools. Text extraction and analysis are processes in the cases where the existing modules failed to predict the permissibility of an email. Moreover, extracted text is accumulated with the signature of the image to make the analysing process easier.

The text categorization technique is analysed by Naïve Bayes text classifier and SVM text categorization tasks. To detect the text based on low-level features like texture, text regions and colour distribution are used. Text extraction by OCR techniques has two issues: one is the computational cost and the other is distorting images in the OCR. The solutions to reduce computational costs are to use a hierarchical architecture for spam filter and image signature which helps to reduce computational complexity. The solutions for distorting images are no content obscuring techniques used by spammers, and CAPTCHA produces high recognition performance using OCR techniques. For execution of text classification method, many measures require to be pursued, like vocabulary building, indexing, classifier preparing and categorization of fresh emails.

Shivananda and Nagabhushan [Shi, 09] suggested a mixture method that integrates linked element study and an unattended threshold for division of text from the intricate backdrop. This suggested method recognizes the applicant text areas grounded on edge detection succeeded by a linked element study. Due to backdrop intricacy, it is in addition feasible that a non-text area may be recognized as a text area. To conquer this issue, texture elements of linked elements ought to be taken out and the element values require to be studied. Eventually, the threshold value for every detected text area is derived mechanically from the info of comparable picture area to carry out foreground division. This recommended method can deal with document images with differing backdrop of multiple colours. In addition, it can deal with foreground text of any colour, font and dimension.

A picture categorization method that takes out elements from actual spam images is suggested by Wu et al. [Wu, 05]. The selected elements are cumulatively calculated on all the pictures enclosed to an email. Elements connected to the existence of text are as follows: the count of detected text areas, the fraction of pictures with detected text areas and the comparative region filled with text (generally indicated as text region). Grounded on the presumption that several spam pictures are banners and computer-generated graphics that are part of advertise, the proportion of the number of banner and of graphic images to the total count of enclosed pictures are in addition employed as elements. Banners are found out by means of their aspect proportion, height and width. Graphics detection is grounded on the presumption that computer-generated graphics generally include homogeneous backdrop and compact texture; it is performed by means of texture study grounded on wavelets. The proportion of external pictures, that is pictures positioned on a remote server and connected in the email's body to the entire count of external and enclosed pictures, is in addition employed as an element, provided that spam emails frequently include connections to external images.

Nevertheless, Aradhye et al. [Ara, 05] assert the elements to be individually calculated for separate picture. Primarily, the text region is taken out by an impromptu procedure. Four colour saturation and heterogeneity elements are then suggested as the colour saturation and heterogeneity values in spam pictures are intermediate betwixt those of valid pictures of natural scenes and valid computer-generated graphics. In this task, a two-class SVM classifier is employed. Lastly, Liu et al. [Liu, 10b] employed elements derived from corner and edge detection to characterize text regions, where colour elements are employed to characterize the graphic components of spam pictures. Specifically, colour difference, common colour coverage, and the count of colours included in a picture are employed as elements (with no definite presumption regarding their dispensation in spam and valid pictures), along with colour saturation components, with similar reasons as in.

2.2.4 Text Extraction

Gupta and Banga [Gup, 12] submitted an approach for taking out text from pictures like document images, scene pictures and so on. The writers have employed discrete wavelet transform (DWT) for taking out text info from intricate pictures. The Sobel edge detector is employed for taking out text edges. There are two dissimilar methods employed for text extraction from intricate pictures, namely area-grounded method and texture-grounded method.

The area-grounded procedure employs the elements of the colour or greyscale in a text area or their dissimilarities about the setting. This approach is fundamentally separated into two subgroups: edge-grounded and connected component (CC)-grounded approaches. The edge-grounded approach chiefly concentrates on the lofty contrast betwixt text and backdrop. CC-grounded approach procedure regards text as a group of individual CCs, each possessing definite force and colour dispensation.

The texture method is grounded on the idea of textural elements. In this approach, Fourier transforms, discrete cosine transform, and wavelet decomposition are usually employed.

Chandrasekaran and Chandrasekaran [Cha, 11] initiated a strong method for text extraction and identification in pictures. Primarily, the input picture is filtered by the Median filter to eliminate any deafening sound. Then, employing LOG edge detector, edges are found out. The morphological dilation surgery is used in object regionalization. All the CCs are then taken out, and all non-text character elements are deserted by a two-measure procedure. Elements are then taken out from the extracted elements. These elements make the element vector for SVM. For identifying separate characters, these elements are tried with SVM. Consequently, all identified characters are combined to make text lines.

Zhang et al. [Zha, 08] suggested a fresh text extraction algorithm under backdrop image grounded on two-dimensional wavelet transforms. For the algorithm, primarily the picture is changed into the wavelet sphere and following that a sliding window is fixed to scan high-frequency sub-bands, by means of calculating the wavelet texture elements of the picture in the sliding window, *k*-sources clustering algorithm is employed to categorize the picture into text region, plain setting region and intricate setting region. Lastly, arithmetical morphology functions are used on the text region to find the text placements precisely.

Hedjam et al. [Hed, 10] suggested the approach of strong partition procedure for text extraction from the historical record pictures. This approach is grounded on Markovian-Bayesian clustering on regional graphs on both pixel and local scales. It comprises of three measures. An over-partitioned map of the input picture is produced in the first measure. The consequent map gives an affluent and precise semi-mosaic portions. In the second measure, the map is processed. Identical and adjacent sub-areas are combined in unison to make precise text forms. The result of the second measure that includes precise forms is processed in the last measure in which employing clustering with firm count of categories, partition will be acquired. The approach uses considerably the regional and spatial relationship and cohesion on both the picture and betwixt the stroke sections, and hence is strong with regard to the disgrace. The consequent partitioned text is even, and feeble links and loops are maintained because of the strong disposition of the approach.

Nirmala and Nagabhushan [Nir, 12] presented an RGB colour paradigm for the contribution of intricate colour document images. An algorithm to find the text areas employing Gabor filters succeeded by extraction of text employing colour quality brightness is in addition created. The method comprises of three phases. On the basis of Gabor elements, the applicant image portions comprising text are found out in phase-1. Due to the intricate setting, some quantity of high-frequency non-text objects in the backdrop is, in addition, found out as text objects in phase-1. Some quantity of fake text objects is discarded by carrying out the related element study in phase-2. In phase-3, the image portions comprising textual data, which are acquired from the preceding phase, are binarized to draw the foreground text. From the input colour document image, the colour quality brightness is withdrawn. Employing this colour element, the threshold value is deduced instinctively. This method deals with both printed and handwritten colour document images with foreground text in any colour, font, size, and orientation.

2.2.5 Other Studies

Experiment by Chen et al. [Che, 17] suggested an extended edition of HOG, Scale and Translation Robust HOG (STRHOG) for enhancing the functioning of clever character identification. Experiments on two public data sets and one of our data sets have exhibited promising outcomes for our task. To filter spam images on the cloud, the enhanced smart character identification is beneficial. Immediate neighbour classifier is employed for the smart character identification in order to make a reasonable comparison with other approaches. It is anticipated that the functioning ought to be furthermore enhanced by employing superior classifiers like an indistinct neural network. Nevertheless, in order to make a reasonable comparison with other approaches, closest neighbour classifier is employed for the smart character identification. It is anticipated that the functioning ought to be furthermore enhanced by employing superior classifiers like an indistinct neural network. Researcher [Dad, 19] discussed popular machine learning approaches towards spam filtering. Study reveals background applications of machine learning techniques in the direction of spam email filtering. The strength and weakness of the existing machine learning approaches were addressed. They suggested deep adversarial learning and deep learning techniques are the future directions to detect spam email.

2.3 OCR Technique

Study by Biggio et al. [Big, 07] recommended a halfway method to identify image spam grounded on finding out the existence of content unclear methods, and explain a feasible execution grounded on two low-stage image elements targeted at finding out unclear methods whose result is to settle the OCR capability leading to character disintegrating or combining, or in the existence of deafening sound impinging with characters in the binarized image. The analysis employed upper-case letters and lower-case letters in English alphabet and regarded four types of degradation: small unsystematic deafening sound elements of various dimensions and thickness, characters broken by a grid composed of one-pixel broad white lines with diverse gaps, diminished character gaps leading to combined characters and characters combined with a grid composed of one-pixel broad black lines with diverse gaps. Data set of 186 actual spam images gathered at the writers' private email accounts. Content hiding methods evidently targeted at vanquishing OCR tools were used by spammers on 96 of these images, whereas the balance 90 images were either tidy or included a restricted quantity of unsystematic deafening sound possibly targeted at conquering disclosure methods grounded on image digital signatures, that is nevertheless insignificant to OCR. Employing Fine Reader 7.0 Professional, binarization of these images was conducted. In spite of the employment of content hiding methods, we observed that on 29 out of 96 deafening sound images, the result of the binarization was a good-quality image. Nevertheless, in certain instances, the recommended steps fell through to find image degradation. Certain images were hidden employing various backdrop colours that led to big sound elements meddling with characters.

Work by Yamakawa and Yoshiura [Yam, 12] analysed Tesseract-OCR, which is an open source OCR software, to use this OCR to find out image spam emails. This document specialized Tesseract-OCR for identification of spam words by creating language info for spam words. This document assessed the identification capability of this specialized Tesseract-OCR by the experiment. Nevertheless, this document creates tutoring images by cautiously choosing images containing several diverse fonts, but further types of junk email images are needed for tutoring to enhance identification capability. In addition, in the experiment, the creating procedure takes the maximum time. Hence, mechanization is needed. Mechanical creation of tutoring images allows creating tutoring images from fresh junk email images by including these tutoring images to Tesseract-OCR immediately following the receipt of fresh junk email images.

2.3.1 Low-Level Image Feature

Work by Shao et al. [Sha, 17b] recommended a mixture spam disclosure approach grounded on amorphous data sets grounded on a union of image and text spam identification methods. Specifically, the previous one is grounded on sparse depiction grounded categorization, which concentrates on the universal and regional image elements, and a dictionary studying method to attain spam and a ham sub-dictionary. On the contrary, the textual study is grounded on semantic characteristics of records to evaluate the degree of spitefulness. In particular, we are capable of differentiating betwixt meta-spam and actual spam. Modern outcomes indicate the precision and ability of our method.

2.3.2 Text Extraction

Work by Fu et al. [Fu, 2000] has studied character extraction, grounded on statistical and constructional elements of grey areas, and suggests an energetic regional difference holding line breadth. Accuracy positioning of character sets is accomplished by using horizontal projection and character layouts of binary images beside horizontal and vertical ways, respectively. Further debated is the approach for partition of characters in binary images that are grounded on projection considering stroke width and character dimensions. A fresh approach for character recognition is examined, that is grounded on compound neural networks. An intricate neural network comprises of two sub-nets, with the first sub-net carrying out self-induction of designs through two-dimensional regionally linked three-order networks, the second subset connecting a regionally linked BP networks carrying out categorization. Strengthened dependability of the network identification by presenting rules for refuting recognition. Twenty-eight images are obtained and handled. Experiment effects prove that all characters in 24 images are exactly recognized and the right recognition rate attains 92.8%. The time spent is below 0.2 s for each image. Nevertheless, one character in two images is mistakenly recognized because these images are partially protected, and extraction of characters in other two images is hard since reflex is additionally persistent. In addition, to attain 100% exactness, it is necessary to check manually.

A study by Al-Duwairi et al. [Duw, 13] which uses Detecting Image Spam Using Image Texture Features recommends an image spam filtering method, known as Image Texture Analysis-Based Image Spam Filtering (ITA-ISF), that utilizes low-level image components for image characterization. We assess the functioning of many machine learning-based classifiers and compare their functioning in filtering image spam grounded on low-level image texture components. C4.5 Decision Tree (DT), Support Vector Machine (SVM), Multilayer Perception (MP), Naïve Bays (NB), Bayesian Network (BN) and Random Forest (RF) are these classifiers. Our empirical analyses grounded on two openly accessible data sets prove that the RF classifier outperforms all other classifiers with average accuracy, recollect, exactness and F-measure of 98.6%.

Work by He et al. [He, 17] suggest the additional structure of profound erudition, a linguistic quality hierarchy, embedded with linguistic decision trees, for spam detection, and verify the result of semantic qualities on the spam detection, embodied by the linguistic quality hierarchy. A case study on the SMS message database from the UCI machine learning storehouse has proved that a linguistic quality hierarchy implanted with linguistic decision trees gives a clear method to thorough examining quality effect on spam detection. This method can not only effectively take on 'curse of dimensionality' in spam detection with huge qualities but also enhance the functioning of spam detection when the semantic qualities are made to an appropriate hierarchy.

Work by Olatunji et al. [Ola, 17] venture has been made at probing how SVM and ELM compared on the exclusive and significant issue of email span detection, that is a categorization issue. The significance of email in this current era cannot be magnified. Therefore, the necessity to quickly and correctly discover and segregate unasked-for emails via spam detection arrangement cannot be magnified further. Practical outcomes from experiments conducted employing extremely famous data set suggested that both methods outdid the best previously published methods on the identical prevailing data set used in this research. Nevertheless, SVM functioned better than ELM on comparison scale grounded on precision. ELM outdid SVM considerably when it comes to speed of functioning.

In an experiment by Roy et al. [Roy, 17], Data is exposed to different kinds of encroachment assaults that may diminish the usefulness of any network or arrangements. Continuously varying and the perplexed quality of trespass endeavours on computer networks cannot be handled IDSs which are presently functional. Recognizing and avoiding such assaults is one of the most difficult undertakings. Profound education is one of the most effectual machine learning methods that is getting accepted of late. This document verifies that possible ability of Deep Neural Network as a classifier for the various kinds of encroachment assaults. A relative analysis has also been conducted with Support Vector Machine (SVM). The empirical outcomes prove that the exactness of encroachment disclosure employing Deep Neural Network is acceptable.

In a research by Kajaree and Behera [Kaj, 17] machine learning algorithm is employed for text interpretation, design identification, and several other commercial aims and has resulted in an independent study attraction in info mining to recognize concealed uniformity or deviations in social info which is spreading by every second. This document concentrates on clarifying the idea and development of Machine Learning, certain well-known Machine Learning algorithms and attempts to compare three most prevalent algorithms grounded on certain fundamental views. The sentiment 140 data set was employed, and functioning of every algorithm by way of grooming time, forecasting time, and exactness of forecasting has been recorded and compared.

Work by Shah and Kumar [Sha, 17a] employed ensemble spam classifier method. Corpus was downloaded from the inbox employing python script. Select python combined unit known as Beautiful Soup to handle our raw emails and classify them as spam and ham. Following preprocessing, we estimated the quality list comprising header info such as Subject, From, To, Reply To, Recipients and so on. Every email matches with one row in the feature list (table). Following preprocessing, the body is stemmed, the word of every email is saved in dumpy array where every row matches with one email and the comprehensive list is known as Word List. Since we possess binomial class values, that is, spam and ham, our classifier has to infer either 0 or 1. If 0, then ham, otherwise spam. The logistic relapse is a decisively statistical probabilistic algorithm that functions on dichotomous values. Hence, we give the values to the classifier and provide the two-class values as 0, 1. The classifier grounded on the estimated frequencies and tutoring info provides the yield. The logistic relapse classifier provides the raw yield (inept or regionalized) so as to make the yield effective; we chart the yield of every word in element vector with a provided class, and this approach is known as one versus rest categorization (OvR). The outcome emerges to be One verses Rest precision: 0.86731, i.e. the original capability is 86% that is superior to other categorizations. We recursively select various features and words from the storehouse to make measured enhancement in our outcome. The enhanced outcome that we obtained is One versus Rest precision: 0.88931, that is 89% exactness. This means that our arrangement has the comprehensive ability of finding out spam as 89%. Since we raise the dimension of the training corpus, the classifier turns increasingly smart. The ability of recommended algorithm is extremely reliant on the genetic algorithm. We employed the profound unit to imitate the genetic algorithm and produce forthcoming generations grounded on fitness purpose. Nevertheless, the genetic algorithm can furthermore be equipped to enhance the precision rate. Work by Zhiwei et al. [Zhi, 17] detects the email spam effectively and efficiently with high accuracy becomes a significant study. In this study, data mining will be utilized to process machine learning by using different classifiers for training and testing and filters for data preprocessing and feature selection. It aims to seek out the optimal hybrid model with higher accuracy or based on other metric's evaluation. The experiment results show accuracy improvement in email spam detection by using hybrid techniques compared to the single classifiers used in this research. The optimal hybrid model provides 93.00% of accuracy and 7.80% false positive rate for email spam detection. To perform the text classification, several machine learning algorithms can be used, such as Naïve Bayes, Decision Tree and Support Vector Machine (SVM). Since the experimental environment may differ, it cannot say that which specific classifier is the best. Different classifiers have its own characteristic and advantages. But, in general, the Decision Tree, SVM and Naïve Bayes are the most welcomed classifiers when process the machine learning for text classification, such as email spam detection. Hybrid techniques are able to perform well by providing better accuracy within different domains compared with a single classifier.

A work by Silva et al. [Sil, 17] has developed the novel hybrid ensemble technique which unites the predictions attained from classifier by utilizing actual text models with its variants generated by implementing semantic indexing and text normalization approaches. This work utilized MDL text, text expansion tool for evaluating blocked ham rate, Matthews's correlation coefficient and spam caught rate.

2.4 Deep Learning Methods for Spam Detection

A deep model [Che, 19] was introduced without back propagation method for spam detection. The authors proved that less training cost was achieved with minimum hyper-parameters using neural network method. The outperformed results were achieved by proposed deep cascade forest method. They also elaborated the working of deep cascade forest method as input document split into words and textual information were extracted as feature vector, then deep neural network learning the text features layer by layer. The output of each base model is based on the prediction of probability of a sentence. In deep learning model, *K*-fold cross-validation is used to reduce the over fitting risk. For experimentation, two publicly available data sets were used to train the model. To calculate training time and testing time, F1-score was analysed. The data set split into 70:30 for training and testing process. Author proved that Naïve Bayes looks appropriate for spam detection among various machine learning approaches as well as to achieve low training cost and high accuracy rate.

A work by Saidani et al. [Sai, 19] has developed a solution to spam filtering problem by two levels text semantic analysis. In the initial level, deep learning approach was used to specific domain by Word2Vec to categorize the email. In the next level, the text from email content was extracted by summarizing the rule to achieve better results based on precision measure. Word embedding by neural network was used in Word2Vec which helps to convert words into equivalent vector in an *N*-dimensional space. In domain-specific spam detection, semantic features were classified using SVM, Naïve Bayes and KNN. For experimentation, two publicly available data sets for spam and ham, i.e. Enron and Ling-spam, respectively, were used. Topic extraction model in specific domain attains efficient spam email detection.

Shahariar et al. [Sha, 19] proposed a model for spam review detection using deep learning method based on labelled and unlabelled data. This model is explained with four phases: preprocessing and data acquisition will be the first phase using Natural Language Processing (NLP) for better performance. Active learning algorithm is used in the second phase to label the data from the unlabelled one. Feature selection process will be the third phase, which uses deep learning approaches for MLP that use TF-IDF (Term Frequency, Inverse Term Frequency), Word2Vec for word embeddings which helps to represent text as numeric values. Various classifiers were used to classify review as ham or spam. For experimentation, Ott data set for labelled and Yelp data set for unlabelled were used. The authors proved that deep learning approach needs more training than machine learning approach.

Jain et al. designed the combined architecture [Jai, 19] for spam detection using deep learning approach using CNN and LSTM. To increase the performance of the proposed model, the authors utilized knowledge-based WordNet and ConceptNet in domain-specific embedding words. For evaluation, sms spam and twitter data set were measured by true positive, true negative, false positive, false negative, precision, recall, F-measure and accuracy. As a result, proposed combined model produced better performance than traditional model. Improved accuracy rate of SMS data set is 1.16%, and improved accuracy rate of Twitter data set is 2.05%.

2.5 Prototypes

2.5.1 HoneySpam

Andreolini et al. [And, 05] proposed a method called Honeypots for fighting spam at the source. The concept submitted with HoneySpam is to fight spammers at the origin rather than the target that stores bandwidth and decreases traffic. HoneySpam is obtained from Honeypots (a machine or arrangement that persuades assailant by feigning as an actual sufferer for collecting the maximum amount of info for forthcoming protection). HoneySpam constructs energetic web pages that are connected together and includes an enormous count of particular legitimate email IDs to reduce the pace of email harvesting (the deed of browsing web pages so as to obtain fresh email IDs to forward spam emails). All these particular email IDs are managed by HoneySpam SMTP (Simple Mail Transfer Protocol) server that tracks spammer activity. Besides, HoneySpam gives false open proxy or relays (a kind of Internet Service that is employed for being obscure) to log spammers' activity, stores network traffic and stops them.

2.5.2 *Phonetic String Matching*

Freschi et al. [Fre, 06] created a spam filtering procedure employing phonetic string matching. Incoming emails go by the following four missions:

1. Normalization: It is the method of mapping graphical indication into initial character. For instance, '*V*' maps into '*v*'.
2. Null-char removal: It is the method used to eliminate worthless non-alphabetic characters. For instance, '-'.
3. Key-specific disambiguation: It is the method used to locate analogous letters in lieu of the balance non-alphabetic characters, identical to normalization method but utilized following the phase of null-char removal.
4. Phonetic transcription: It is the method of searching the matching system of phonetic symbols that depict identical English pronunciation. For instance, 'Buy' and 'Bye'.

The outcome of every work is forwarded to imprecise string matching (the method of penetration, cancellation and exchanging of nearby characters that returns most apt connected strings to initial string) which leads to four flags. These four flags are joined in principle combination unit (the method of flags ultimately uniting) to label as spam or ham.

2.5.3 *ProMail*

Tseng et al. [Tse, 07] developed an advanced email social network called ProMail for spam detection. ProMail is an email social network grounded on anti-spam procedure appropriate for handling enormous count of emails. By employing email server logs, ProMail constructs graph model that includes nodes (for every email ID) and edges (for every email communication). It has a gradually modernized outline that appends fresh email IDs and eliminates obsolete emails from the graph model. Following that, ProMail uses a ranking procedure for rating every node, and in place of rating or re-rating every node it rates a subset of nodes in the graph model to quicken the procedure of spam detection. ProMail differentiates spam email account from ham (non-spam) account by catching sight of every email ID score in the graph model; those above the threshold are regarded as spam email IDs.

2.5.4 Zombie-Based Approach

Lieven et al. [Lie, 07] suggested a spam filtering procedure grounded on recurring designs. Currently, spammers employ zombie machines or spam-bots to forward email messages. SMTP protocol carried out inside zombie machine does not correctly pursue the SMTP benchmark. For instance, zombie machines do not stay for distribution of email message or retain the emails in a heap till transmitted. By analysing the conduct of an SMTP server and comparing it with a normal SMTP protocol, the system differentiates spam SMTP senders from real senders. For discerning spam conduct, upon getting a fresh connection, the system instinctively discards connection and halts for the sender's reconnection appeal. If the sender reconnects, the system will label it as ham as it is responding in conformity with the SMTP protocol, that is, the SMTP server ought to resend failed-delivery-emails. On the contrary, if the sender did not reconnect or carry out another action, relying on the kind of action, the system would label sender as spam or give an extra chance for the sender to reconnect.

2.5.5 SMTP Logs Mining Approach

Lam and Yeung [Lam, 07] submitted a learning method for spam detection grounded on social networks. Writers in this document have suggested a machine learning anti-spam procedure that analyses the conduct of an email server so as to differentiate legal user conduct from spam user conduct. The system constructs a graph model grounded on SMTP server logs. Every node depicts an email account and edges on the graph display relative email connection among email accounts. Eliciting and examining seven elements from each email account, the system allots a score to every node that would be employed to categorize spam email server from ham email sender.

2.6 Previous Works

A study by Sanches and Moreira [San, 17] recommended an image anti-spam system that utilizes diverse approaches of image element extraction and a fake neural model to categorize emails. The extraction procedures are assessed both separately and in the mixture. The neural model is thoroughly assessed employing openly accessible databases. The employment of these databases is explained elaborately so as to ease replicating outcomes. In addition to studying the categorization skill of the recommended arrangement, this analysis further assesses its computational costs, along with costs for extracting elements and categorizing images. The outcomes are encouraging both in terms of rates of right categorization and of fake positives created by the anti-spam system, along with by way of its computational cost.

Work by Saidani et al. [Sai, 17] recommended a method for email spam detection grounded on text semantic study at two stages. The first stage permits classification of emails by particular areas (for instance, health, education, finance and so on). The second stage employs semantic elements for spam detection in each particular territory. We demonstrate that the recommended procedure gives an effective portrayal of the inner semantic framework of email content that permits for further exact and explainable spam filtering outcomes compared to present procedures.

2.6.1 Integrated Approach

Work by Kumar and Biswas [Kum, 17a] declared Image spam as a kind of email spam in which the textual message is implanted within an image submitting it as a picture. This document suggests a Support Vector Machine (SVM) with Gaussian kernel grounded classifier to discern the spam. In our experiment, we have employed openly accessible data sets with SVM and Gaussian kernel grounded classifier proving that our way provides fine functioning over studied classifiers for measurement of F-measure, recollect, precision and accuracy.

What are the pitfalls in the existing methods?

Author & Year	Method/Algorithm	Drawback
Dredze et al. [Dre, 07]	Feature-based classification	Feature extraction can be much more time-consuming, especially when features depend on complex image.
Biggio et al. [Big, 11]	SVM	The drawback of the SVM is their relatively complex training and categorizing algorithms and also the high time and memory consumptions during training stage and classifying stage.
Chandrasekaran and Chandrasekaran [Cha, 11]	SVM	The training time can be very large if there are large numbers of training examples and execution can be slow for nonlinear SVMs
Das and Prasad [Das, 14]	OCR, SVM	High false positive rate
Syal and Garg [Sya, 14]	DWT, gradient method and SVM	

2.7 Research Gap

The examination of the previous literature revealed the following limitations in previous researchers:

The advent of image spam challenges the traditional text-based spam filters to handle texts embedded in images. Hence there is a need for a technique that could appropriately handle text embedded image spam. In this regard, character segmentation and recognition techniques should be proposed as a system which should appropriately identify an image as Ham or Spam. The Optical Character Recognition-based image spam filtering occupies more memory and consumes time, hence not suitable for online spam detection on server-side. Most image spam techniques analyse image contents, through embedded text detection, colour analysis or embedded text extraction which are slow as images are far larger than most emails. Some of the most indicative properties of spam images such as the existence of text regions still need to be enhanced with respect to performance and robustness against image variation. The image spam filtering methods tend to have high false positive rates in labelling a mail ham as spam. The privacy issues of anti-spam filtering let the collection of the negative set so difficult which leads to the increase in false positive rate.

The present research identified two main challenges within the text recognition and segmentation procedures which have led to the aforementioned limitations. Firstly, the integration of different steps involved in text recognition and segmentation increases the complexity of the system. Secondly, integration of different procedures increases error. Hence, the present research recognized the various techniques used for character recognition and segmentation and examined the ways of appropriately integrating different techniques used. Discrete wavelet transforms, and Hough transforms are used in combination as a hybrid character segmentation technique. Discrete wavelet transform (DWT) applies the high-low sub-band 2D Haar DWT feature in a twofold manner so as to increase vertical edge recognition and decrease background noise in applications. Secondly, the use of Hough transform addresses the edge point grouping for detecting shapes such as eclipse, straight line and circle. Template matching technique and Contour analysis facilitate the recognition of characters. The integration of the aforementioned character segmentation and character recognition techniques enables a system that could integrate text segmentation to the core and may tend to address the limitations stated in the previous researchers. Furthermore, the segmented characters are feature extracted, and the extracted patterns are analysed to examine whether an image is Ham or Spam. Combining all the aforementioned techniques is the novelty of the present research. In addition, new techniques such as spatial frequency cross-correlation, improved local binary pattern, nearest neighbour search and Optimization Consideration were also performed to improve the efficiency of image spam detection.

2.8 Summary

The main objective of this chapter is to review the recent research works conducted on the text and image spam email classification. This chapter presented a detailed review on spamming with its related terminologies. Questions such as how spams are hazardous to email users along are elucidated with the solutions suggested using classification algorithms. The chapter also explains the different types of email filtering methods that are used in the past. There are various machine learning algorithms that are discussed in this study. All algorithms discussed have both advantages and disadvantages. This chapter also explains the techniques used for the image email classification. The various techniques of text extraction and segmentation method are also discussed in this chapter. Finally, the chapter concludes with the issues of all the related works which supported the researcher to present a novel algorithm by identifying the research gap.

References

[Alm, 11] Almeida, T. A., Almeida, J., & Yamakami, A. (2011). Spam filtering: How the dimensionality reduction affects the accuracy of naive Bayes classifiers. *Journal of Internet Services and Applications, 1*(3), 183–200. Retrieved from http://www.springerlink.com/index/10.1007/s13174-010-0014-7.

[Awa, 11a] Awad, W. (2011a). Machine learning methods for spam E-mail classification. *International Journal of Computer Science and Information Technology, 3*(1), 173–184. Retrieved from http://www.airccse.org/journal/jcsit/0211ijcsit12.pdf.

[Awa, 11b] Awad, W. A., & ELseuofi, S. M. (2011b). Machine learning methods for E-mail classification. *International Journal of Computer Applications, 16*(1), 39–45. Retrieved from https://pdfs.semanticscholar.org/e011/3ba1c6a737f685f84639645ce9b58b8ca41e.pdf.

[Vij, 18] Vijayasekaran, G., & Rosi, S. (2018). Spam and email detection in big data platform using Naives Bayesian classifier. *International Journal of Computer Science and Mobile Computing, 7*(4), 53–58. Retrieved from https://www.academia.edu/36435374/pdf.

[Ger, 17] Gerardnico. (2017). *Machine learning—K-nearest neighbors (KNN) algorithm—instance based learning*. [Online]. Retrieved from https://gerardnico.com/wiki/data_mining/knn.

[Pat, 13b] Patidar, V. (2013b). A survey on machine learning methods in spam filtering. *International Journal of Advanced Research in Computer Science and Software Engineering, 3*(10), 964–972. Retrieved from https://pdfs.semanticscholar.org/7102/37d884e1fee9ab06897b081ac7ac45643096.pdf.

[Kho, 07] Khorsi, A. (2007). An overview of content-based spam filtering technique. *Academic Journal, 31*(3), 269–277. Retrieved from http://connection.ebscohost.com/c/articles/27465123/overview-content-based-spam-filtering-techniques.

[Meh, 17] Mehdy, M. M., Ng, P. Y., Shair, E. F., Saleh, N. I. M., & Gomes, C. (2017). Artificial neural networks in image processing for early detection of breast cancer. *Computational and Mathematical Methods in Medicine, 2017*(1), 1–15. Retrieved from https://www.hindawi.com/journals/cmmm/2017/2610628/.

[Mar, 09] Marsono, M. N., El-Kharashi, M. W., & Gebali, F. (2009). Targeting spam control on middleboxes: Spam detection based on layer-3 e-mail content classification. *Computer Networks, 53*(6), 835–848. Retrieved from http://linkinghub.elsevier.com/retrieve/pii/S1389128608003988.

[Car, 06] Carpinteiro, O. A. S., Lima, I., Assis, J. M. C., de Souza, A. C. Z., Moreira, E. M., & Pinheiro, C. A. M. (2006). A neural model in anti-spam systems. *Lecture Notes in Computer Science, 4132*(1), 847–855. Retrieved from https://pdfs.semanticscholar.org/534e/acfab6a2e-86146f9a4d5702a4021b2f5a86f.pdf.

[Tia, 12] Tian, Y., Shi, Y., & Liu, X. (2012). Recent advances on support vector machines research. *Technological and Economic Development of Economy, 18*(1), 5–33. Retrieved from http://www.tandfonline.com/doi/abs/10.3846/20294913.2012.661205.

[Say, 11] El-Sayed, H., & El-Bassiouny, N. (2011). *An outlook into the academic-practitioner divide. German University in Cairo. With Implications For International Marketing Education*. https://www.researchgate.net/publication/261393406

[And, 17] Anderson, C., Figa-Saldana, J., Wilson, J. J. W., & Ticconi, F. (2017). Validation and cross-validation methods for ASCAT. *IEEE Journal of Selected Topics in Applied Earth Observations and Remote Sensing, 10*(5), 2232–2239. Retrieved from http://ieeexplore.ieee.org/document/7809062/.

[Sar, 12] Saruladha, K., & Sasireka, L. (2012). A survey of text classification algorithms. *Computer Science and Engineering, 3*(1), 163–222. Retrieved from http://link.springer.com/chapter/10.1007/978-1-4614-3223-4_6.

[Chr, 10] Christina, V., Karpagavalli, S., & Suganya, G. (2010). Email spam filtering using supervised machine learning techniques. *International Journal on Computer Science and Engineering, 2*(9), 3126–3129. Retrieved from http://www.enggjournals.com/ijcse/doc/IJCSE10-02-09-151.pdf.

[Mah, 13] Mahmood Ali, M., Qaseem, M. S., Rajamani, L., & Govardhan, A. (2013). Extracting useful rules through improved decision tree induction using information entropy. *International Journal of Information Sciences and Techniques, 3*(1), 27–41. Retrieved from http://www.airccse.org/journal/IS/papers/3113ijist03.pdf.

[Kum, 17b] Kumar, R., Suresh, K., Ramakrishna, S., & Padmavathamma, M. (2017b). Development of data mining system to compute the performance of improved random tree and J48 classification tree learning algorithms. In *International conference on innovative applications in engineering and information technology (ICIAEIT-2017)* (pp. 128–132). New York: IEEE.

[Wri, 17] Wright, A., Walker, J. P., Robertson, D. E., & Pauwels, V. R. N. (2017). A comparison of the discrete cosine and wavelet transforms for hydrologic model input data reduction. *Hydrology and Earth System Sciences Discussions, 21*, 3827–3838. Retrieved from http://www.hydrol-earth-syst-sci-discuss.net/hess-2017-26/.

[Sya, 14] Syal, N., & Garg, N. K. (2014). Text extraction in images using DWT, gradient method and SVM classifier. *International Journal of Emerging Technology and Advanced Engineering, 4*(6), 477–481. Retrieved from https://pdfs.semanticscholar.org/b909/b943041fb372f2b-5b865abd73679ae2e7a8b.pdf.

[Sah, 10] Saha, S., Basu, S., Nasipuri, M., & Basu, D. K. (2010). A Hough transform based technique for text segmentation. *Journal of Computing, 2*(2), 135–141. Retrieved from http://arxiv.org/abs/1002.4048.

[Gur, 13] Gurov, I. P., Potapov, A. S., Scherbakov, O. V., & Zhdanov, I. N. (2013). Hough and Fourier transforms in the task of text lines detection. In *QCAV2013—11th International conference on quality control by artificial vision. [Online]* (pp. 222–227). Saint Petersburg, Russia: Russia. Retrieved from https://www.researchgate.net/publication/272825230_Hough_and_Fourier_Transforms_in_the_Task_of_Text_Lines_Detection.

[Aye, 17] Ayesh, M., Mohammad, K., Qaroush, A., Agaian, S., & Washha, M. (2017). A robust line segmentation algorithm for Arabic printed text with diacritics. *Electronic Imaging, 2017*(13), 42–47. Retrieved from http://www.ingentaconnect.com/content/10.2352/ISSN.2470-1173.2017.13.IPAS-204.

[Isl, 16] Islam, N., Islam, Z., & Noor, N. (2016). A survey on optical character recognition system. *Journal of Information & Communication Technology-JICT, 10*(2), 1–4. Retrieved from https://arxiv.org/ftp/arxiv/papers/1710/1710.05703.pdf.

[Ali, 15] Ali, N., Isheawy, M., & Hasan, H. (2015). Optical character recognition (OCR) system. *IOSR Journal of Computer Engineering Ver. II, 17*(2), 2278–2661. Retrieved from www.iosr-journals.org.

[Pan, 16] Panchal, T., Patel, H., & Panchal, A. (2016). License plate detection using Harris corner and character segmentation by integrated approach from an image. *International Conference on Communication, Computing and Virtualization, 79*(1), 419–425. Retrieved from http://www.academia.edu/24314198/License_Plate_Detection_using_Harris_Corner_and_Character_Segmentation_by_Integrated_Approach_from_an_Image.

[Jav, 13] Javed, M., Nagabhushan, P., & Chaudhuri, B. B. (2013). Extraction of projection profile, run-histogram and entropy features straight from run-length compressed text-documents. In *2013 2nd IAPR Asian conference on pattern recognition. [Online]* (pp. 813–817). New York: IEEE. Retrieved from http://ieeexplore.ieee.org/lpdocs/epic03/wrapper.htm?arnumber=6778437.

[Rod, 01] Rodrigues, R. J., Vianna, G. K., & Thomé, A. C. G. (2001). *Character feature extraction using polygonal projection sweep (contour detection). [Online]* (pp. 687–695). Berlin: Springer. Retrieved from http://link.springer.com/10.1007/3-540-45723-2_83.

[Pat, 13a] Patel, P. & Tiwari, S. (2013a). Text segmentation from images. International Journal of Computer Applications 67(19), pp. 25–28. Retrieved from https://pdfs.semanticscholar.org/76b0/d192a9ffc9796836b96275ddcadb00f91981.pdf.

[Raj, 16] Rajalingam, M., & Sumari, P. (2016). An enhanced character segmentation and extraction method in image-based email detection. *International Journal of Control Theory and Applications, 9*(26), 171–179.

[Kar, 14] Karanje, U. B., & Dagade, R. (2014). Survey on text detection, segmentation and recognition from a natural scene images. *International Journal of Computers and Applications, 108*(13), 975–8887.

[Kap, 11] Kapoor, R., Gupta, S., & Sharma, C. M. (2011). Multi-font/size character recognition and document scanning. *International Journal of Computer Applications, 23*(1), 21–24. Retrieved from https://pdfs.semanticscholar.org/7f0f/ed910ec1cb0f2704b2a807335d9c365fd98d.pdf.

[Sha, 14] Al-Shatnawi, A. M. (2014). A skew detection and correction technique for Arabic script text-line based on subwords bounding. In *2014 IEEE International conference on computational intelligence and computing research. [Online]* (pp. 1–5). New York: IEEE. Retrieved from http://ieeexplore.ieee.org/document/7238501/.

[Sta, 15] Stahlberg, F., & Vogel, S. (2015). Detecting dense foreground stripes in Arabic handwriting for accurate baseline positioning. In *2015 13th International conference on document analysis and recognition (ICDAR). [Online]* (pp. 361–365). New York: IEEE. Retrieved from http://ieeexplore.ieee.org/document/7333784/.

[Kor, 12] Korchagin, D. (2012). Automatic time skew detection and correction. *International Journal of Computer and Electrical Engineering, 4*, 684–687. Retrieved from http://www.ijcee.org/show-46-687-1.html.

[Tan, 13] Tanase, M. C., Zaharescu, M., & Bucur, I. (2013). Upsampling-Downsampling image reconstruction system. *Journal of Information Systems & Operations Management (JISOM). The Proceedings of Journal ISOM, 7*(2), 294–299. Retrieved from http://jisom.rau.ro/downloads/JISOM-10-2-dec-2016.pdf.

[Sam, 06] Samosseiko, D., & Thomas, R. (2006). The game Goes on: An analysis of modern spam techniques. In: *VB Conference*. 2006, sophos.

[Fum, 06] Fumera, G., Pillai, I., & Roli, F. (2006). Spam filtering based on the analysis of text information embedded into images. *Journal of Machine Learning Research, 7*(1), 2699–2720. Retrieved from http://www.jmlr.org/papers/volume7/fumera06a/fumera06a.pdf.

[Itt, 95] Ittner, D.J., David, D., Lewis Y. D., & Ahn, Z. (1995). Text categorization of low quality images. In: *Proceedings of SDAIR-95, 4th Annual Symposium on Document Analysis and Information Retrieval*. 1995, citeseerx.

[Shi, 09] Shivananda, N., & Nagabhushan, P. (2009). Separation of foreground text from complex background in color document images. In *2009 Seventh international conference on advances in pattern recognition* (pp. 306–309). New York: IEEE.

[Wu, 05] Wu, C.-T., Cheng, K.-T., Zhu, Q., & Wu, Y.-L. (2005). Using visual features for anti-spam filtering. In *IEEE International conference on image processing 2005* (p. III-509). New York: IEEE.

[Ara, 05] Aradhye, H. B., Myers, G. K., & Herson, J. A. (2005). Image analysis for efficient categorization of image-based spam e-mail. In *Eighth international conference on document analysis and recognition (ICDAR'05)* (Vol. 2, pp. 914–918). New York: IEEE.

[Liu, 10b] Liu, Y.-N., Huang, C.-H., & Chao, W.-L. (2010b). *Image segmentation based on the normalized cut framework.*

[Gup, 12] Gupta, N., & Banga, V. K. (2012). Localization of text in complex images using Haar wavelet transform. *International Journal of Innovative Technology and Exploring Engineering., 1*(1), 1–111.

[Cha, 11] Chandrasekaran, R., & Chandrasekaran, R. M. (2011). Morphology based text extraction in images. *International Journal of Computer Science and Technology., 2*(4), 103–107.

[Zha, 08] Zhang, X. W., Zheng, X. B., & Weng, Z. J. (2008). Text extraction algorithm under background image using wavelet transforms. In *Proceedings of international conference on wavelet analysis and pattern recognition* (pp. 30–31). Hong Kong: IEEE.

[Hed, 10] Hedjam, R., Farrahi Moghaddam, R., & Cheriet, M. (2010). Text extraction from degraded document images. In *2010 2nd European workshop on visual information processing (EUVIP)* (pp. 247–252). New York: IEEE.

[Nir, 12] Nirmala, S., & Nagabhushan, P. (2012). *Foreground text segmentation in complex color document images using Gabor filters* (Vol. 6, p. 669). Berlin, Germany: Springer.

[Che, 17] Chen, J., Zhao, H., Yang, J., Zhang, J., Li, T., & Wang, K. (2017). An intelligent character recognition method to filter spam images on cloud. *Soft Computing, 21*(3), 753–763. Retrieved from http://link.springer.com/10.1007/s00500-015-1811-5.

[Dad, 19] Dada, E. G., Bassi, J. S., Chiroma, H., Abdulhamid, S. M., Adetunmbi, A. O., & Ajibuwa, O. E. (2019). Machine learning for email spam filtering: Review, approaches and open research problems. *Heliyon, 5*(6), e01802. https://doi.org/10.1016/j.heliyon.2019.e01.

[Big, 07] Biggio, B., Fumera, G., Pillai, I., & Roli, F. (2007). Image spam filtering using visual information. In *14th International conference on image analysis and processing (ICIAP 2007). [Online]* (pp. 105–110). New York: IEEE. Retrieved from http://ieeexplore.ieee.org/document/4362765/.

[Yam, 12] Yamakawa, D., & Yoshiura, N. (2012). Applying tesseract-OCR to detection of image spam mails. In *2012 14th Asia-pacific network operations and management symposium (APNOMS). [Online]* (pp. 1–4). New York: IEEE. Retrieved from http://ieeexplore.ieee.org/document/6356068/.

[Sha, 17b] Shao, Y., Trovati, M., Shi, Q., Angelopoulou, O., Asimakopoulou, E., & Bessis, N. (2017b). A hybrid spam detection method based on unstructured datasets. *Soft Computing, 21*(1), 233–243. Retrieved from http://link.springer.com/10.1007/s00500-015-1959-z.

[Fu, 2000] Fu, Z., Bian, F., Zhou, Z., & Hu, Q. (2000). Algorithm for fast detection and identification of characters in gray-level images. *International Archives of Photogrammetry and Remote Sensing, 33*(Part B3), 305–311. Retrieved from http://www.isprs.org/proceedings/xxxiii/congress/part3/305_xxxiii-part3.pdf.

[Duw, 13] Al-Duwairi, B., Khater, I., & Al-Jarrah, O. (2013). Detecting image spam using image texture features. *International Journal for Information Security Research, 3*(4), 344–353. Retrieved from http://infonomics-society.org/wp-content/uploads/ijisr/published-papers/volume-3-2013/Detecting-Image-Spam-Using-Image-Texture-Features.pdf.

[He, 17] He, H., Watson, T., Maple, C., Mehnen, J., & Tiwari, A. (2017). A new semantic attribute deep learning with a linguistic attribute hierarchy for spam detection. *Proceedings of the International Joint Conference on Neural Networks, 2017*, 3862–3869.

[Ola, 17] Olatunji, S. O. (2017). Extreme learning machines and support vector machines models for email spam detection. In *2017 IEEE 30th Canadian conference on electrical and computer engineering (CCECE). [Online]* (pp. 1–6). New York: IEEE. Retrieved from http://ieeexplore.ieee.org/document/7946806/.

[Roy, 17] Roy, S. S., Mallik, A., Gulati, R., Obaidat, M. S., & Krishna, P. V. (2017). *A deep learning based artificial neural network approach for intrusion detection [Online]* (pp. 44–53). Retrieved from http://link.springer.com/10.1007/978-981-10-4642-1_5.

[Kaj, 17] Kajaree, D., & Behera, R. (2017). A survey on machine learning: Concept, algorithms and applications. *International Journal of Innovative Research in Computer and Communication Engineering, 5*(2), 1302–1309.

[Sha, 17a] Shah, N. F., & Kumar, P. (2017a). *A comparative analysis of various spam classifications. [Online]* (pp. 265–271). Berlin: Springer. Retrieved from http://link.springer.com/10.1007/978-981-10-3376-6_29.

[Zhi, 17] Zhiwei, M., Singh, M. M., & Zaaba, Z. F. (2017). Email spam detection: A method of metaclassifiers stacking. *Proceedings of the 6th International Conference on Computing & Informatics, 200*(200), 750–757. Retrieved from http://icoci.cms.net.my/PROCEEDINGS/2017/Pdf_Version_Chap16e/PID200-750-757e.pdf.

[Sil, 17] Silva, R. M., Alberto, T. C., Almeida, T. A., & Yamakami, A. (2017). Towards filtering undesired short text messages using an online learning approach with semantic indexing. *Expert Systems with Applications, 83*, 314–325. Retrieved from http://linkinghub.elsevier.com/retrieve/pii/S0957417417303056.

[And, 05] Andreolini, M., Bulgarelli, A., Colajanni, M., & Mazzoni, F. (2005). HoneySpam: Honeypots fighting spam at the source. In *Proceedings USENIX steps to reducing unwanted traffic*. Boston: Usenix.

[Fre, 06] Freschi, V., Seraghiti, A., & Bogliolo, A. (2006). Filtering obfuscated email spam by means of phonetic string matching. In *Proceeding ECIR'06 Proceedings of the 28th european conference on advances in information retrieval*. Berlin, Heidelberg: Springer-Verlag.

[Tse, 07] Tseng, C.-Y., Huang, J.-W., & Chen, M.-S. (2007). ProMail: Using progressive email social network for spam detection. In *Advances in knowledge discovery and data mining* (pp. 833–840). Berlin, Heidelberg: Springer.

[Lie, 07] Lieven, P., Scheuermann, B., Stini, M., & Mauve, M. (2007). Filtering spam email based on retry patterns. In *IEEE international conference on communications (ICC'07)*. Glasgow, Scotland: Glasgow Caledonian University.

[Lam, 07] Lam, H. Y., & Yeung, D. Y. (2007). A learning approach to spam detection based on social networks. In *Proceedings of the fourth conference on email and anti-spam (CEAS). [Online]*. USA: CRC Press. Retrieved from https://books.google.co.in/books?id=plnvwf4PVDkC&pg=PA142&lpg=PA142&dq=A+Learning+Approach+to+Spam+Detection+based+on+Social+Networks.

[San, 17] Sanches, B. C., & Moreira, E. M. (2017). Detecting image spam with an artificial neural model. *International Journal of Computer Science and Information Security, 15*(1), 296–315.

[Sai, 17] Saidani, N., Adi, K., & Allili, M. S. (2017). *A supervised approach for spam detection using text-based semantic representation. [Online]* (pp. 136–148). Berlin: Springer. Retrieved from http://link.springer.com/10.1007/978-3-319-59041-7_8.

[Kum, 17a] Kumar, P., & Biswas, M. (2017a). SVM with Gaussian kernel-based image spam detection on textual features. In *2017 3rd International conference on computational intelligence & communication technology (CICT). [Online]* (pp. 1–6). New York: IEEE. Retrieved from http://ieeexplore.ieee.org/document/7977283/.

[Che, 19] Chen, K., Zou, X., Chen, X., & Wang, H. (2019). An automated online spam detector based on deep cascade forest. In *The second annual international conference on science of cyber security (SciSec 2019). [Online]* (pp. 33–46). Berlin: Springer. Retrieved from https://link.springer.com/chapter/10.1007/978-3-030-34637-9_3.

[Sai, 19] Saidani, N., Adi, K., & Allili, M. S. (2019). Semantic representation based on deep learning for spam detection. In *12th international symposium on foundations and practice of security (FPS 2019). [Online]* (pp. 72–81). Berlin: Springer. Retrieved from https://link.springer.com/chapter/10.1007/978-3-030-45371-8_5.

[Sha, 19] Shahariar, G. M., Biswas, S., Omar, F., Shah, F. M., & Hassan, S. B. (2019). *Spam review detection using deep learning* (pp. 0027–0033). New York: IEEE. Retrieved from https://sci-hub.tw/https://ieeexplore.ieee.org/document/8936148.

[Jai, 19] Jain, G., Sharma, M., & Agarwal, B. (2019). Spam detection in social media using convolutioal and long short term memory neural network. *Annals of Mathematics and Artificial Intelligence, 85*, 21. https://doi.org/10.1007/s10472-018-9612-z.

[Das, 14] Das, M., & Prasad, V. (2014). Analysis of an image spam in email based on content analysis. *International Journal on Natural Language Computing, 3*(3), 129–140. Retrieved from http://www.airccse.org/journal/ijnlc/papers/3314ijnlc13.pdf.

[Dre, 07] Dredze, M., Gevaryahu, R., & Elias-Bachrach, A. (2007). Learning fast classifiers for image Spa. In *Proceedings of the conference on email and anti-spam*. New Delhi: CEAS.

[Big, 11] Biggio, B., Fumera, G., Pillai, I., & Roli, F. (2011). A survey and experimental evaluation of image spam filtering techniques. *Pattern Recognition Letters, 32*(10), 1436–1446. Retrieved from http://linkinghub.elsevier.com/retrieve/pii/S0167865511000936.

Chapter 3
Methodology

3.1 Introduction

Over the past years, several methods emerged for the detection of image spam wherein these researchers were the result of the changes in the ways through which spammers embed spam content in email images. While this has become a threat for the internet community, researchers have come up with novel techniques which have to some extent supported the ways of mitigating image-based spam in emails [Big, 11]. In the present section, the researcher examined previous researchers and the current methods for the detection of image spam.

The emergence of distributed detection approaches paved the way to solving complex image spam detection within distributed systems wherein the same is found to act as a scalable method. It was [Zho, 03] who utilized the feature signatures for the detection of spam in distributed systems wherein the same tends to identify never seen spam and fuzzy images. It was [Gup, 16; Jin, 10; Kur, 15] who proposed the reputation-based approaches that could detect spam images through filtering. However, the increase in communication overhead acts as the limitation for such a method. These researchers have a single limitation; the basis for high storage overhead is evident which makes the spam detection processes practicably impossible and complicated.

Previous studies have proposed several techniques for the detection of spam images wherein the premise is on the use of pattern recognition techniques. For instance, the probabilistic decision tree method is developed by Gao et al. [Gao, 08] by applying colour histogram and the gradient orientation histograms for the detection of spam. Chen et al. [Che, 17] utilized the various STRHOG methods for the recognition of characters in an intelligent manner wherein the provisioning of such a system is made in the cloud, and cloud-based images are filtered. However, Chowdhury et al. [Cho, 15] proposed the spam filtering method which utilizes the BPNN classifier wherein the file features of the image which is spam and is of low

M. Rajalingam, *Text Segmentation and Recognition for Enhanced Image Spam Detection*, EAI/Springer Innovations in Communication and Computing,
https://doi.org/10.1007/978-3-030-53047-1_3

level are extracted, and the visual feature points were used for the detection of spam. In addition, research by Wakade et al. [Wak, 13] introduced the method of image spam detection wherein the paper utilized the J48 algorithms and various features for the identification of spam images. The various features that were used in the previous paper are as follows: a number of colours, luminance, white pixel concentration, colour saturation, deviation of hue and colours and so on. The findings of the previous research revealed that the pattern recognition techniques are deemed to be effective for the detection of spam images [Bis, 06]. These techniques do not involve the requirements such as storage space for the storing of identifiers for all spam items identified as a state in [Zho, 03].

For the effective detection of spam, two different categories of spam detection were performed—low- and high-level feature extraction, and classification and image spam detection. Conventional studies including [Che, 10; Gao, 10; Bha, 08; Wan, 10] all utilized low-level feature extraction techniques. The research by Bhaskar et al. [Bha, 08] presented two methods for the detection of image spam wherein the first method uses colour, texture and shape as visual features and classification was performed using SVM. Such a classification yielded 95% accuracy in all the cases selected. The second method, however, utilizes the near-duplicate detection of images wherein the method utilizes the clustering of image Gaussian Mixture Models (GMM) which is based on the principle of Agglomerative Information Bottleneck (AIB), and has high accuracy to around 93% during the prediction of spam.

Binary Filtering with Multi-Label Classification (BFMLC) was proposed by Chen et al. [Che, 10] for the detection of both spam image filtering and user preferences. The method comprises of the classification phase which is two-staged, user-oriented multi-label classification and filter-oriented binary classification. The experimental findings of the research revealed that recognition of spam images was done with maximum accuracy wherein spam images are classified based on predefined topics. A feature extraction scheme was proposed by Wang et al. [Wan, 10] on low-level visual as well as metadata features wherein the same comprises of features such as image height, width, file type and so on; however, the visual features comprise of variance, number of colours, primary colour and colour saturation which is characterized through histogram. An accuracy of 95% was found in the previous research.

Studies concerned with high-level image characteristics contain file format, file name, aspect ratio, image area, horizontal and vertical resolution and other data. The research by Krasser et al. [Kra, 07] devised the frame structure to feature extract and classify the images in a rapid manner based on obtained features. Four basic image features were utilized such as height, width, the file type of the image, size of the file and so on which could be derived from an image in a rapid manner, and the cost of computation is relatively low. C4.5 algorithm and SVM algorithm was used for the building of decision tree and SVM, respectively. However, [Uem, 08] framed a technique that accesses the conventional Bayesian filter to observe the image data including the file size, name, image area and compressibility. Moreover, it is also implemented in GIF images which create a lot of image spam.

3.2 Proposed Design

For the present research, three tasks are performed (CSRC framework) wherein each task is based on the steps that are performed in the research for the detection of image spam. The step by step procedure includes the use of different algorithms which could be used in combination to acquire better image spam detection accuracy. In the present research, the researcher utilized the processes of segmentation, recognition and image spam detection using the methods of DWT, Hough transforms, spatial frequency cross-correlation for automatic segmentation, contour analysis, improved local binary pattern for text recognition, and the extraction of visual features using SVM and KNN classifiers. For the present research, the researcher utilized the Dredze data set which is a vital source for both image spam and ham emails data set adopted from [Dre, 07]. Figure 3.1 depicts the processes performed in the present research.

The initial task begins with the processes of character segmentation wherein DWT and Hough transforms will be used for the segmentation of characters from images wherein the researcher identified the basis for spatial frequency cross-correlation for automatic segmentation. The data set used for the first task is the Image Spam data set acquired from [Dre, 07]. The metrics used are theta, peaks and rho of edge map image wherein pixel count analysis is used to improve the accuracy of segmenting characters from the image.

Secondly, task 2 (Column 2) depicts the processes of character recognition wherein the initial stage begins with the survey of previous literature to understand the concepts of character recognition which led the researcher to understand the benefits of using Template matching, Contour Analysis, Bounding box as a combined method. In addition, the improved local binary pattern for text recognition has been identified to act as a better method towards improving the performance of the character recognition methods used. The data set used for the task is once again the Image Spam data set acquired from [Dre, 07]. Finally, the third task implies the use of visual feature extraction using SVM and KNN classifiers wherein the image is checked whether it falls under Ham or Spam.

3.2.1 Data Set

For the present research, the Dredze image spam data set is used which is compiled by Dredze et al. [Dre, 07]. The image spam data set contains 2173 images in Spam Archive corpus, 2359 images in personal ham corpus and 1248 images in personal spam. The main motivation of using the aforementioned data set is due to ease of collecting both image ham and spam emails from a single repository. Furthermore, the data set selected contains different types of spam and ham images. The Dredze data set is an open source image ham and spam data set which has several image spam and ham. The collection of the images for both ham and spam was performed

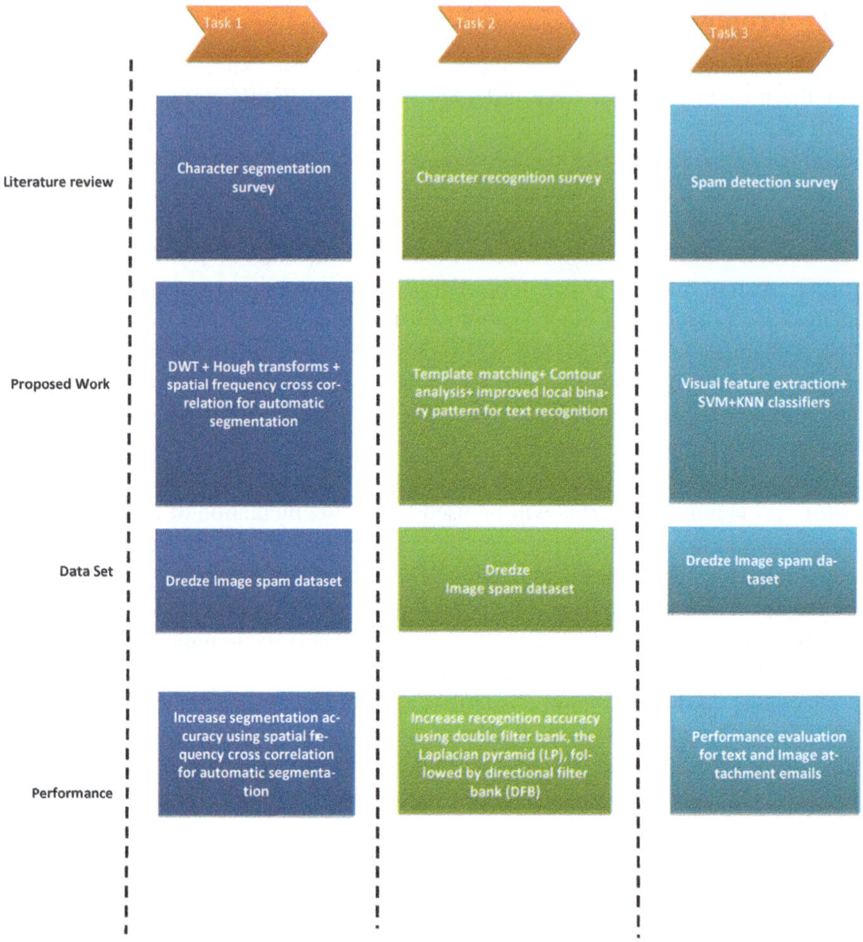

Fig. 3.1 Architecture of the proposed image spam detection system

by identifying images from the mailboxes of real users by the previous researcher (Mark Dredze). The first use of the Dredze data set was recorded in the year 2006 by Fumera et al. [Fum, 06]. The image spam data set contains only files in image format (such as .jpeg and .gif); however, in general, image spam come with different formats such as .jpeg, .gif, .bmp and .png. The format named 'BMP' or bitmap image comprises of raster images. GIF is also a format similar to Bitmap; however, it supports bits per pixel and supports animations. However, PNG is developed to improve GIF format and employs lossless data compression. However, the present research considers both GIF and JPEG images.

The present research considered the image spam data set that [Dre, 07] used for the evaluation of the three proposed tasks: character segmentation, character recognition and spam email detection. The image spam data set collected from [Dre, 07] is chosen due to the evidence that there exist different kinds of image spam messages which are generated by spammers for the sake of escaping from spam filters. These image spam messages are sufficient and robust for the segmentation and recognition evaluation with regard to spam detection domain [Duw, 13]. The data set consists of 2173 images in the SpamArchive corpus, 2359 images in personal ham corpus and 1248 images in personal spam. There are 'text only', 'randomized' and 'wild background' images. Text only images contain only texts, whereas images with randomization are added with random colour pixels, stripes and colour shades. The last type is an image with wild background wherein the images are embedded with a noisy background.

However, there are other types of images which are even more appealing for users. They include animated gif, multipart Images, and standard images which are attractive and are least filtered by spam filters. Figure 3.2 shows image spam with text only, and Fig. 3.3 shows image spam with randomisation.

3.2.2 *Corpus*

Corpus refers to the collection of images used in the present research. All the images are collected from the Dredze image spam data set. The images are collected from the data set (both Ham and Spam) and are used for the research. All the images in the Dredze data set are considered for training, and 25 images in both spam and ham are considered as testing images.

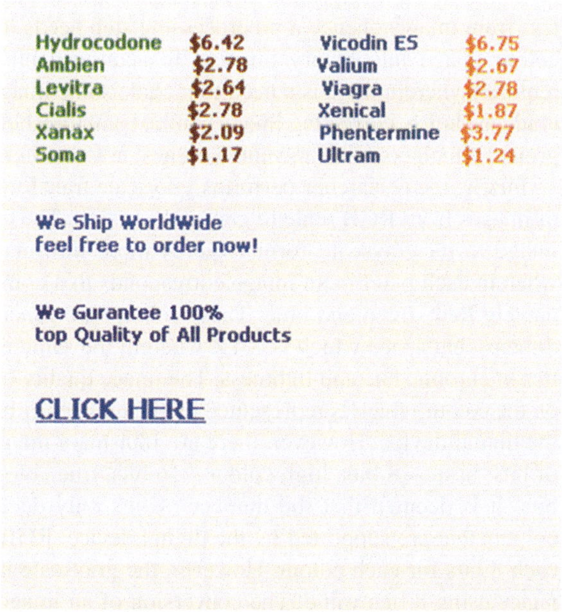

Fig. 3.2 Image spam with text only

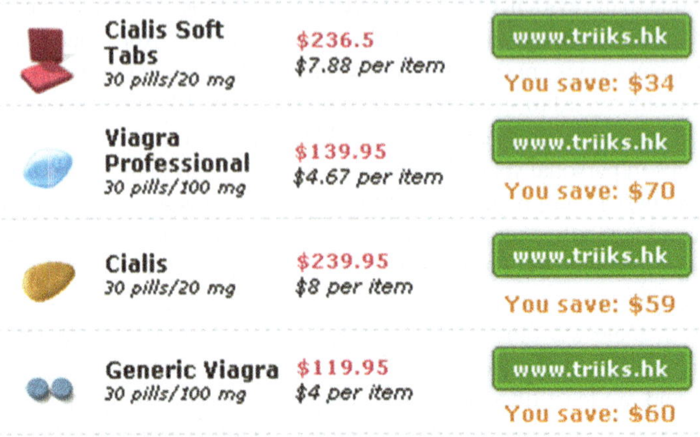

Fig. 3.3 Image spam with randomisation

3.2.3 *Preprocessing*

In the present research, the collected images from the image spam data set need to be preprocessed to improve the image spam detection process. In general, preprocessing needs to be performed on any image in any level such as colour, greyscale or a binary document image which covers both graphics and text. This step is important to improve the efficiency of character recognition since the processing of colour images has high computational overheads. Such colour images contain watermarks or non-uniform background which makes it challenging to acquire the document text from images; hence a preprocessing step needs to be performed to convert the colour image into binary images. To acquire a binary image, several steps are required wherein there is a need to include some image enhancement techniques to eradicate noise, correcting image contrast and thresholding for the removal of background which contains unwanted scenes, noise, watermarks and so on.

Firstly, the researcher performs grayscale transformation for the transformation of images from RGB scale to greyscale images. The conversion of a coloured RGB image to its grayscale form requires more information about the colour image wherein each pixel in an image corresponds to a combination of the three colours, namely Red, green and blue. These colours are generally represented in the three-dimensional space which is XYZ wherein the same is depicted by the characteristics of chroma, hue and lightness. The image quality of a colour image is dependent on the colour which is represented by the number of bits that could be supported by the digital device. However, there are four main image types based on the number of bits: basic—8 bits, high colour—16 bits, true colour—24 bits, deep colour—32 bits. It is deemed that the number of bits only decide the maximum number of colours that are supported by the digital device. RGB image occupies 24 bits with each 8 bits for each colour. However, the grayscale image is represented by luminance using 8 bits value. The conversion of an image from grayscale into RGB is simply depicted as the conversion of 24 bit into 8-bit grayscale value [Sar, 10]. In the present research, the colour image is converted into a grayscale image using the RGB value approximation using luminance RGB components added with the value of chrominance which provides good-quality grayscale images.

The next step in preprocessing is the normalization phase which is a scaling technique [Pat, 15]. The normalization of images is a process that transforms the pixel intensity value range. The applications of normalization include the transformation of poor contrast photographs wherein the same is sometimes known as histogram stretching or contrast stretching [Gon, 04]. The next step after normalization is filtering and reduction of noise which is performed using noise-cleaning methods [Ver, 17].

3.3 Experimental Set-Up and Performance Evaluation

The proposed algorithms that will improve the detection accuracy of text and image-based email classification are experimented to achieve the research objectives. All the proposed algorithms are fulfilled using MATLAB (version R 2013a), and the experimentations are performed on an Intel(R) Core (TM) i5 machine with a speed of 2.60 GHz and 8.0 GB RAM using Windows 8.1 64-bit Operating System. For experimentation, images were taken from image spam data set which facilitates fast classification and hence is utilized in the study [Das, 14]. The proposed approaches (Character segmentation using DWT, Hough transforms and spatial frequency cross-correction for automatic segmentation; character recognition using template matching, contour analysis and local binary pattern for text recognition, and visual feature extraction using SVM and KNN classifiers) are evaluated for performance wherein the entire Dredze data set images for training are used to measure accuracy by parameters such as true positive, true negative, false positive, false negative, precision, recall and F-measure.

3.3.1 Performance Evaluation Measures—Character Segmentation and Recognition

Performance efficiency of the proposed algorithms is assessed through the results with respect to detection accuracy. Furthermore, comparisons among the proposed algorithms and current state-of-the-art methods are shown in the experimentations. The accuracy of proposed algorithms is measured using True Positive (TP), True Negative (TN), False Positive (FP), False Negative (FN), Recall and Precision.

Image spam data set is downloaded from [Dre, 98]. The image spam data set contains 2173 images in the SpamArchive corpus, 2359 images in personal ham corpus and 1248 images in personal spam. For the assessment of the proposed algorithms with respect to the detection a ccuracy, the proposed algorithms are compared with the existing individual methods.

Furthermore, to examine the performance of the proposed approaches (Hybrid Character segmentation, Template matching and contour analysis, and shape-based feature extraction), the values of false negative, false positive, true negative, true positive, precision, recall and F-measure are measured and are compared with the values of the factors acquired in previous researchers. Following is the description of the performance analysis indicators used in the present research:

3.3.1.1 False Positive Rate (FP)

When a test reports that a message being sent is spam though the reality is the message is ham. This is identified through the equation,

$$\mathrm{FP} = \frac{b}{b+d} \tag{3.1}$$

3.3.1.2 False Negative Rate (FN)

When a test reports that a message being sent is a ham though the reality is the message is spam. This is identified through the equation,

$$\mathrm{FN} = \frac{c}{c+a} \tag{3.2}$$

3.3.1.3 Precision Rate (P)

The rate of retrieved instances which are relevant is known as the Precision rate (P). P is calculated by the ratio of accurate segmented characters to a total of accurate segmented characters and false positive.

$$\mathrm{P} = \frac{\text{Correctly segmented characters}}{\text{Correctly segmented characters} + \text{False positive}} \tag{3.3}$$

3.3.1.4 Recall (R)

The rate of a fraction of relevant instances that are retrieved is called as Recall (R). R is calculated by the ratio of correctly predicted characters to the sum of correctly identified characters plus false negative.

$$R = \frac{\text{Correctly segmented characters}}{\text{Correctly segmented characters} + \text{False negative}} \tag{3.4}$$

3.3.1.5 F-Measure (F)

A system's F-measure is defined as the weighted harmonic mean of its precision and recall, and F-Measure (F) is calculated using precision and recall.

$$F = 2 * \frac{P * R}{P + R} \tag{3.5}$$

3.3.1.6 Accuracy (A)

Overall accuracy is calculated as a weighted arithmetic mean of Precision and Inverse Precision or Recall and Inverse Recall which is given by,

$$\text{Accuracy}(A) = \frac{TP + TN}{TP + TN + FP + FN} \tag{3.6}$$

3.4 Summary

This chapter covesrs the information about the data sets and the evaluation procedures used in the research. The chapter covered information regarding the steps involved in preprocessing and the experimental set-up used by the researcher towards the development of the image ham spam detection system.

References

[Big, 11] Biggio, B., Fumera, G., Pillai, I., & Roli, F. (2011). A survey and experimental evaluation of image spam filtering techniques. *Pattern Recognition Letters, 32*(10), 1436–1446. Retrieved from http://linkinghub.elsevier.com/retrieve/pii/S0167865511000936.

[Zho, 03] Zhou, F., Zhuang, L., Zhao, B. Y., Huang, L., Joseph, A. D., & Kubiatowicz, J. (2003). Approximate object location and spam filtering on peer-to-peer systems. In *Proceedings of the ACM/IFIP/USENIX 2003 International Conference on Middleware* (pp. 1–20). Rio de Janeiro, Brazil: U. C. Berkeley. Retrieved from https://dl.acm.org/citation.cfm?id=1515915.1515917.

[Gup, 16] Gupta, R., Singha, N., & Singh, Y. N. (2016). Reputation based probabilistic resource allocation for avoiding free riding and formation of common interest groups in unstructured P2P networks. *Peer-to-Peer Networking and Applications, 9*(6), 1101–1113. Retrieved from http://link.springer.com/10.1007/s12083-015-0389-0.

[Jin, 10] Jin, X., & Chan, S.-H. G. (2010). Detecting malicious nodes in peer-to-peer streaming by peer-based monitoring. *ACM Transactions on Multimedia Computing, Communications, and Applications, 6*(2), 1–18. Retrieved from http://portal.acm.org/citation.cfm?doid=1671962.1671965.

[Kur, 15] Kurdi, H. (2015). HonestPeer: An enhanced EigenTrust algorithm for reputation management in fP2Pg systems. *The Journal of King Saud University Computer and Information Sciences, 27*(3), 315–322. Retrieved from https://slideheaven.com/distributed-classification-for-image-spam-detection.html.

[Gao, 08] Gao, Y., Yang, M., Zhao, X., Pardo, B., Wu, Y., Pappas, T. N., & Choudhary, A. (2008). Image spam hunter. In *2008 IEEE International Conference on Acoustics, Speech and Signal Processing, March 2008* (pp. 1765–1768). New York: IEEE.

[Che, 17] Chen, J., Zhao, H., Yang, J., Zhang, J., Li, T., & Wang, K. (2017). An intelligent character recognition method to filter spam images on cloud. *Soft Computing, 21*(3), 753–763. Retrieved from http://link.springer.com/10.1007/s00500-015-1811-5.

[Cho, 15] Chowdhury, M., Gao, J., & Chowdhury, M. (2015). Image spam classification using neural network. In B. Thuraisingham, X. Wang, & V. Yegneswaran (Eds.), *Security and privacy in communication network* (pp. 622–632). Cham, USA: Springer. Retrieved from http://link.springer.com/10.1007/978-3-319-28865-9_41.

[Wak, 13] Wakade, S., Liszka, K. J., & Chan, C.-C. (2013). Application of learning algorithms to image spam evolution. In S. Ramanna, L. C. Jain, & R. J. Howlett (Eds.), *Emerging paradigms in machine learning* (pp. 471–495). Berlin, Heidelberg: Springer. Retrieved from http://link.springer.com/10.1007/978-3-642-28699-5_18.

[Bis, 06] Bishop, C.M. (2006). *Pattern recognition and machine learning*. [online]. Retrieved from http://www.library.wisc.edu/selectedtocs/bg0137.pdf.

[Che, 10] Cheng, H., Qin, Z., Fu, C., & Wang, Y. (2010). *A novel spam image filtering framework with multi-label classification*. pp. 282–285.

[Gao, 10] Gao, Y., Choudhary, A., & Hua, G. (2010). A comprehensive approach to image spam detection: From server to client solution. *IEEE Transactions on Information Forensics and Security, 5*(4), 826–836. Retrieved from http://ieeexplore.ieee.org/document/5585752/.

[Bha, 08] Bhaskar, M., Saurabh, N., Manish, G., & Wolfgang, N. (2008). Detecting image spam using visual features and near duplicate detection. In *Proceedings of the 17th international conference on World WideWeb Beijing. [Online]*. China: ACM. Retrieved from https://arxiv.org/ftp/arxiv/papers/1212/1212.1763.pdf.

[Wan, 10] Wang, C., Zhang, F., Li, F., & Liu, Q. (2010). Image spam classification based on low-level image features. In *2010 International Conference on Communications, Circuits and Systems (ICCCAS). [online]* (pp. 290–293). Berkeley: IEEE. Retrieved from http://ieeexplore.ieee.org/document/5581998/.

[Kra, 07] Krasser, S., Tang, Y., Gould, J., Alperovitch, D., & Judge, P. (2007). Identifying image spam based on header and file properties using C4.5 decision trees and support vector machine learning. In *2007 IEEE SMC information assurance and security workshop* (pp. 255–261). Berkeley: IEEE.

[Uem, 08] Uemura, M., & Tabata, T. (2008). Design and evaluation of a Bayesian-filter-based image spam filtering method. In *2008 International Conference on Information Security and Assurance (ISA 2008)* (pp. 46–51). Berkeley: IEEE.

[Dre, 07] Dredze, M., Gevaryahu, R., & Elias-Bachrach, A. (2007). Learning fast classifiers for image Spa. In *Proceedings of the conference on email and anti-spam*. New Delhi: CEAS.

[Fum, 06] Fumera, G., Pillai, I., & Roli, F. (2006). Spam filtering based on the analysis of text information embedded into images. *Journal of Machine Learning Research, 7*(1), 2699–2720. Retrieved from http://www.jmlr.org/papers/volume7/fumera06a/fumera06a.pdf.

[Duw, 13] Al-Duwairi, B., Khater, I., & Al-Jarrah, O. (2013). Detecting image spam using image texture features. *International Journal for Information Security Research, 3*(4), 344–353. Retrieved from http://infonomics-society.org/wp-content/uploads/ijisr/published-papers/volume-3-2013/Detecting-Image-Spam-Using-Image-Texture-Features.pdf.

[Sar, 10] Saravanan, C. (2010). Color image to grayscale image conversion. In *2010 Second International Conference on Computer Engineering and Applications. [online]* (pp. 196–199). Indonesia: IEEE. Retrieved from http://ieeexplore.ieee.org/document/5445596/.

[Pat, 15] Patro, S.G.K. & Sahu, K.K. (2015). *Normalization: A Preprocessing Stage.* [online]. 2015. Arxiv. Retrieved December 25, 2017, from https://arxiv.org/ftp/arxiv/papers/1503/1503.06462.pdf.

[Gon, 04] Gonzalez, R. C., & Woods, R. E. (2004). *Digital image processing* (2nd ed.). Gainesville, FL: Prentice Hall Publication.

[Ver, 17] Verne, J. (n.d.). *Image Pre-Processing*. [online]. Semantic scholar. Retrieved December 25, 2017, from https://pdfs.semanticscholar.org/cc43/a71e05cfc49ab0777b82ca94d181f779149f.pdf.

[Das, 14] Das, M., & Prasad, V. (2014). Analysis of an image spam in email based on content analysis. *International Journal on Natural Language Computing, 3*(3), 129–140. Retrieved from http://www.airccse.org/journal/ijnlc/papers/3314ijnlc13.pdf.

[Dre, 98] Dredze, M., & Elias-bachrach, A. (1998). Learning fast classifiers for image spam. *Corpus*. [online]. (C). pp. 2005–2008. Retrieved from http://www.cs.jhu.edu/~mdredze/publications/image_spam_ceas07.pdf.

Chapter 4
Character Segmentation

4.1 Introduction

In the present chapter, a hybrid character segmentation algorithm is proposed which combines several aspects such as DWT, Hough transform and pixel count analysis. Based on the examination of previous researchers, the aforementioned techniques were selected and were hypothesized to improve the segmentation efficiency. The present chapter covers the following sections—Sect. 4.2 covers the proposed hybrid character segmentation algorithm, Sect. 4.3 elaborates on the experimental results and analysis and Sect. 4.4 covers the summary.

4.2 Proposed Hybrid-Based Character Segmentation

The process of character segmentation involves the division of a digital image into several distinct segments called blocks. Every block will comprise a single character which is known as a token. Character segmentation has acquired momentum in the recent years owing to the developments in OCR techniques. An example segmentation of an image into blocks is depicted in Fig. 4.1. The purpose of character segmentation is to segment the regions with characters which are further fed for the character recognition phase.

The proposed digital image character segmentation is shown in Fig. 4.2. The proposed method consists of two main components: (1) preprocessing component and (2) segmentation component. The preprocessing component prepares the image into a simplified form for the next segmentation component. The preprocessing component performs three main tasks which include RGB to greyscale conversion, binary conversion and removal of connected components. The character segmentation component performs three main tasks: application of DWT, line segmentation

M. Rajalingam, *Text Segmentation and Recognition for Enhanced Image Spam Detection*, EAI/Springer Innovations in Communication and Computing,
https://doi.org/10.1007/978-3-030-53047-1_4

Fig. 4.1 Image taken from a data set with segmented characters

and finally character segmentation. However, to improve the accuracy of character segmentation, the researcher utilized pixel count analysis and spatial frequency cross-correlation. The combination of all these techniques enables good segmentation of the characters from the image.

4.2.1 RGB to Greyscale

In this procedure, the colour image (RGB format) is transformed into a greyscale image. The process of conversion of a colour image to a greyscale image demonstrates its luminance, wherein the values of its Red, Green and Blue (RGB) primaries in linear intensity are obtained by encoding using the technique of gamma expansion. 30% red value, 59% green value and 11% blue value are increased together; these weights are based on accurate RGB primaries selection. Irrespective of the scale including 0.0–1.0, 0–255 and 0–100%, the subsequent number is the preferred linear luminance value; Moreover, it required gamma compressed to retrieve to traditional greyscale demonstration. Using human eye approach,

$$\text{Grey} = 0.299 * \text{Red} + 0.587 * \text{Green} + 0.114 * \text{Blue} \tag{4.1}$$

Three major colours red, green and blue to the greyscale value are fixed to adjust the varied gamma based on our requirement.

4.2.2 Binarization and Removal of Connected Components

Binarization is the method of renovating the RGB and greyscale images to the black and white pixel images. Only with the black and white images, noise could be detached competently without moving the character pixels. This greyscale is converted into the binary image using the Otsu's method [Ots, 79]. Otsu's thresholding method comprises repeating through all the possible threshold values and analysing

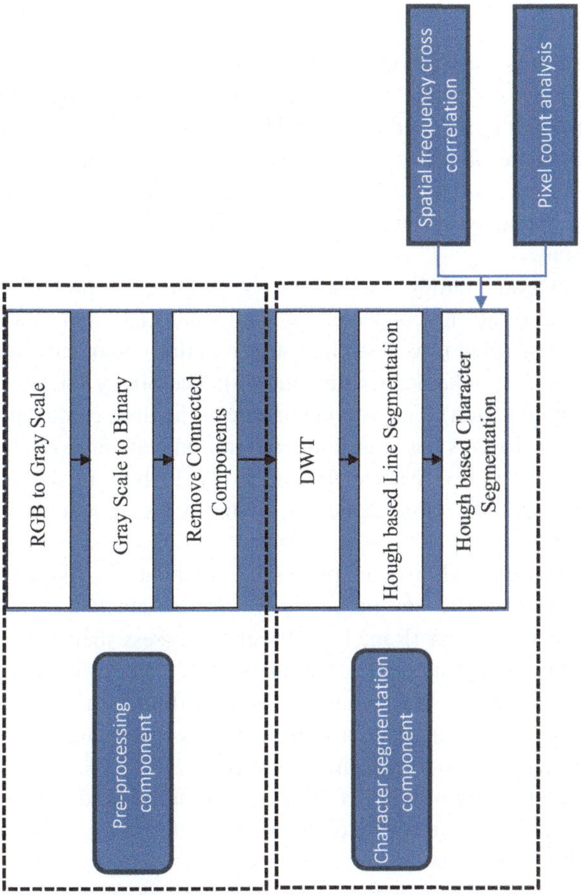

Fig. 4.2 Flow diagram of character segmentation

the amount of the spread for the pixel levels at each side of the threshold such as the pixels (foreground or the background). The objective is to identify the threshold value in which the sum of the foreground and background spreads is least. With the process of binarization complete, all the connected components (objects) are eliminated which have fewer than 15 pixels from the binary image.

4.2.3 Discrete Wavelet Transform (DWT)

Image segmentation is used to locate the boundaries (lines, curves) and objects in the images. Similarly, segmentation of image is the process of converting the pixel of an image which in turn shares the identical label with some visual characteristics. Image segmentation results included the overall image or contours set extracted from the image. Every pixel that exists in the image region has some similar properties including the intensity, texture or colour and are significantly different regarding the same features.

Images are segmented into line and character for the given preprocessing which is after the removal of all the connected components of an input image. First, the image has to be segmented row-wise (line segmentation), then each row has to be segmented column-wise (character segmentation). The binary image is segmented row-wise (line segmentation). The resultant image of the line segmentation is shown in Fig. 4.7. The transform is based on limited duration, called wavelets, of varying frequency and small waves. Wavelet transform (WT) bids in denoising, object detection, feature analysis, edge detection, connected components removal and image compression.

DWT is widely employed in signal processing applications, like the elimination of deafening sound in audio, the imitation of wireless antenna dispensation and video and audio compression [Ran, 15]. Wavelets possess their strength concentrated on time and are appropriate for the analysis of time difference and fleeting signals. Although most real-life signals are time differing in character, Wavelet transform ensembles many applications that yield positive outcomes. Here, DWT has employed to part the character from the image.

Discrete wavelet transform (DWT) is used to eliminate linked elements. The processed image is once again contributed to the discrete wavelet transform unit to obtain three types of edges and texture as neglected by the conventional edge detectors. Discrete wavelet transform is then used to the preprocessed binary image. The binary image is then disintegrated into one stage via DWT. The DWT disintegrates the signal into four diverse elements of frequency areas like LL, LH, HL, HH elements. For each element in the disintegrated image, the edges are recognized employing Canny Edge Detector Algorithm. With DWT, the detected peripheries turn further accurate and obvious. The primary reason of using wavelet transform for periphery detection is that it can eliminate the deafening sound while traditional periphery operators recognize deafening sound pixels as periphery pixels. This area describes discrete wavelet transform and in addition explains character partition through

DWT. DWT is a cautiously planned mathematical instrument for hierarchically disintegrating an image. DWT is the multi-resolution interpretation of an image, and the deciphering can be handled continually from a near-to-the-ground settlement to the combined settlement [Jef, 12]. The DWT divides the signal into high- and low-frequency sections. The high-frequency section consists of info regarding the peripheral elements, whereas the low-frequency section is separated further into high- and low-frequency sections.

The gradual procedure in two-dimensional DWT, for each stage of disintegration, principally executed DWT in a perpendicular way, succeeded by the DWT in a parallel way. Following the first stage of disintegration, there are four sub-bands: LL1, LH1, HL1 and HH1. For every subsequent stage of disintegration, the LL sub-band of the previous stage is employed as the intake.

To carry out second-stage disintegration, the DWT is used to LL1 that disintegrates the LL1 band into the four sub-bands LL2, LH2, HL2 and HH2. To carry out third-stage disintegration, the DWT is used to an LL2 band that disintegrates this band into the four sub-bands—LL3, LH3, HL3 and HH3. This leads to ten sub-bands per element. LH1, HL1 and HH1 encircle the maximum frequency bands started in the image tile, whereas LL3 encircles the minimum frequency band. The procedure of three-tier disintegration is illustrated in Fig. 4.3; the instance of three-stage disintegration, and origin taken from [Kas, 12].

The image is composed of pixels which are arranged in a two-dimensional matrix, to every pixel indicates the digital comparable of image force. In special area nearby pixel, values are extremely connected and therefore unnecessary. As per instruction to contract images, these repetitions present among pixels require being eliminated. DWT processor alters the special area pixels into frequency area information that is depicted in frequency points, compound sub-bands and exhibiting diverse timescale. One of the obvious elements with the use of the 2D-DWT is the alteration of the image specimens into a further compressible structure.

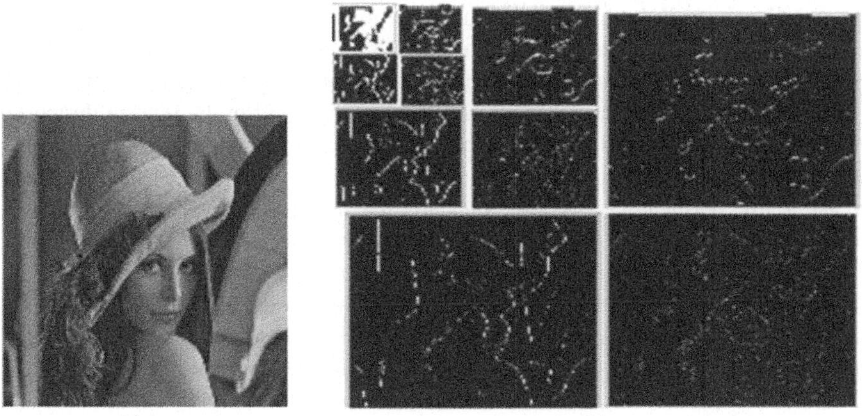

(a) (b)

Fig. 4.3 (**a**) Original image and (**b**) three-level DWT decomposition

4.2.4 Hough-Based Line and Character Segmentation

In an automated study of digital images, a general issue is observing the modest shapes like circle, ellipse or straight line. In all the instances, a periphery detector can be employed as a preprocessing stage to accomplish image pixels or image points that are on the selected curve in the image space. However, because of faults in one or the other image data or the periphery detector, they may be separated, wanting or dissociate pixels or points on the selected curves along with special digressions among the ellipse or lines or circle and the deafening sound periphery points as acquired from the periphery detector. For these particulars, it is often necessary to set the extracted periphery elements to an appropriate group of ellipses or lines or circles.

Hough transform is chiefly employed in many applications. It has the ability to recognize a wide scope of forms like arcs, straight lines, even arbitrary forms in image processing and computer vision. A variety of expansions of Hough transform has been investigated and operational in various domains. The Hough transform has been widely employed in radar detection, medical imaging applications and is beneficial in agriculture to track down fruits [Tu, 15]. In the current study, Hough transform is employed to partition the lines and characters from the image email. The employment of the Hough transform is to discourse of the periphery detection issue by producing it feasible to attain gatherings of peripheral points into an object by conducting an obvious voting procedure over a group of parameterized image objects. The measures entailed in Hough transform is shortly explained in the following.

Let us contemplate on its separated margin point (x, y) in the image level. There could be an unrestricted number of lines that could permeate this point. Each of these lines can be regarded as an explanation to the certain particular equation. In the modest structure, a line can be declared in the slope-intercept structure as $y = mx + c$, where m is the slope of the line with respect to x-axis, and c is the intercept on y-axis made by the line. Any line can be classified by these two criteria pair (m, c). Because all the lines that permeate a given point (x, y), there is a peculiar value of c structure, given by $c = y - mx$. The group of (m, c) values is identical to the lines permeating point (x, y) form a line in (m, c) span. Each point in image space (x, y) matches to a line in restricted area (m, c) and in the reverse direction, each point in (m, c) area imitates to a line in image area (x, y). The Hough transforms function by permitting each element point (x, y) vote in (m, c) space for each feasible line permeating it. These votes are matched in a storage battery. Presume that a particular (m, c) has one vote, this means that there is an element point over and done with permeating lines. If it has two votes, then it means two element point's deception on that line. If a place (m, c) in the storage battery has n votes, this means that 'n' feature point's deception on that line. Hough transform method is used to divide lines and to discover the horizontal lines in input image. For every line, all the lines are extracted from image to recognize the position of the character, and the character from image is partitioned.

4.2.4.1 Hough Transform Algorithm [Sah, 10]

1. Acquisition of all the preferred feature points in the image space
2. For each feature point in image space
    (a) For each option in the accumulator that passes through the feature point
    (b) Increment the position in the accumulator
    (c) Acquisition of local maxima in the accumulator
    (d) On prerequisite plot each maximum in the accumulator back to image space

4.2.5 Spatial Frequency Correlation

In the field of spatial frequency or 2D filtering, the key component is correlation filtering process. It helps to identify, correct and translate the input image simply using training images since correlation filtering consists of shift invariance. Correlation filters can generate correlation peaks, and thus the recognition of the objects needs not to be centralized in the analysis part [Fer, 15]. The correlation output is based on the peaks at the same time the presence of the object examined through the relative heights of the correlation peaks. Furthermore, the position of the peaks is used to find out the position of the object.

Integration operation is the basis for the correlation filtering. Thus, it helps to obtain the input images with minimal damage thereby increases quality of the output. Furthermore, correlation filtering has tolerance towards noise, high refinement quality and provides closed-form expressions [Kum, 02]. Fast Fourier Transforms (FFTs) are used in the cross-correlation which provides speed and quality in output. This method especially recognizes an object from input image with respect to a filter or template and training sets updated as new images are attained rather than retaining this filter from scratch. Additional advantages of implementing this correlation-based technique areas data mining process are impossible due to the usage of image pixels.

Matched filter (MF) is the most fundamental correlation filter. It helps to detect additive white noise corrupted reference images. However, the performance of MF lacks at the time of the distortions in the reference image, for example, rotations and scale changes.

Distance classifier correlation filters (DCCF) is a type of correlation filter which is created according to the idea of variation in the distance of a training image and test image [Mah, 96]. Within a dimensional space (d^2), all fingerprint images appear as a point wherein all axes represent various image pixels. After that, the training fingerprint image from a person indicates a little number of points within the image space. An example of DCCF method is shown in Fig. 4.4 with three types including m_1, m_2 and m_3. Transform H is measured followed by within a d^2 dimensional space the training sets m_1, m_2 and m_3 are maximally separated. DCCF method with three classes is shown in Fig. 4.4.

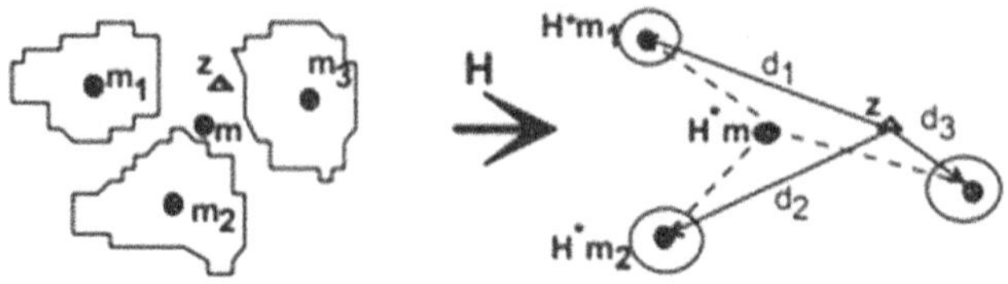

Fig. 4.4 DCCF method with three classes (Source: Adopted from [Kum, 02])

If a test input (z) seems then, it is analysed to the same transform H and its distances to all three types in the transformed field are projected. In line with that, the class with the smallest distance is utilized for the input assignment.

The application of correlation filters tools is crucial in signal processing and pattern recognition like biometric analysis [Kum, 06; Tho, 07], action recognition [Rod, 08], object alignment [Bod, 13], object tracking [Hen, 12; Bol, 10], object [Rod, 13; Hen, 13; Bod, 14], and videos-based event retrieval [Rev, 13] as they are capable of identifying and categorizing numerous objects in a single act at the same time [Fer, 15]. In the present research, spatial frequency correlation is used with Hough's transforms for improved segmentation accuracy.

4.2.6 *Overall Hybrid Algorithm*

Character segmentation from image is a daunting task since it suffers from characteristics such as occluded character and low-resolution image. The difficulty is to segment a character when detecting white space between words. When the characters are touched or overlapped, projection profile method does not give good results, therefore, the following hybrid algorithm is used. At each input image point, a number of lines are plotted at different angles. The vertical projection is used to separate the base characters using the white space between them. The hybrid method of DWT and Hough transform to segment characters from image email improves the performance of character segmentation.

4.2.6.1 Hybrid Algorithm

Input: After the process of binarization, the preprocessed image is provided as input to the hybrid algorithm. The following steps depict how the discrete wavelet transform and Hough transform operate with the selected image:

1. Apply DWT—Discrete wavelet transform is applied to the preprocessed binary image.
2. The binary image is decomposed into single level through DWT.
3. The LL, LH, HL, HH components are identified.
4. For each component in the decomposed image,

   4.1 Find the Edges by using Canny Edge Detector Algorithm [Can, 86]
5. End For.
6. Hough transform technique is applied to segment lines.
7. Find the horizontal lines in image.
8. Extract all the lines from image.
9. For each line in the image,
   9.1 Identify the character location.
   9.2 Apply pixel count analysis.
   9.3 Apply spatial frequency cross-correlation.
   9.4 Segment the character.
10. End For.

Output: Line and Character Segmented Image.

4.3 Experimental Results and Analysis

4.3.1 Experimental Set-Up

For experimentation, 23 images were tested and were taken from the image spam data set [Dre, 07]. Figure 4.5 shows sample image from image spam data set. These images are in two groups that are 11 Spam and 12 Ham images. For the purpose of identifying both ham and spam images from the selected images of the Dredze data set, both ham and spam images are taken and are displayed in Fig. 4.4 for the sake of notifying which images fall in the specific categories. In the present research, all images were taken for training and 23 images for testing. The training images comprise several images with different image characteristics. This is in line with the testing images wherein each image has its own characteristics when it comes to noise, background and so on. Furthermore, the use of 23 samples for testing is in line with the research by De Campos [Cam, 09] which used only 15 images for both training and testing, however, revealed better results in terms of performance. All these images are selected from the data set based on the various kinds of images present in the data set, the uniqueness in each image and to show the efficiency of the proposed ham spam detection method.

Fig. 4.5 Sample ham and spam images considered for illustration

4.3.2 Experimental Task

There are two main experimental tasks: preprocessing and the segmentation. The first experimental task is a preprocessing task which converts a given image into greyscale and then into binary.

The second task is the process of character segmentation which is used to convert given image greyscale and greyscale to binary. After the process of binarization, the preprocessed image is given as an input to the DWT algorithm to remove the connected components which eliminate all the connected components (objects) lesser than 15 pixels from the binary image. DWT removes noise and detects the edges of the text with clarity and accuracy. The detected text edge points are practised to groupings into an object by carrying out a voting method. Hough transforms is applied initially to segment horizontal lines, secondly to identify character location, and finally to segment the character.

4.3.3 Results of Preprocessing Component

The result of preprocessing is shown in Fig. 4.6.

The images as greyscale output are then binarized. The output binary image has values of 0 (black) for all pixels in the input image with luminance less than level and 1 (white) for all other pixels. The value of level is in the range of [0, 1]. Following binarization, the line and connected components were removed and then is segmented using DWT and Hough transforms.

Fig. 4.6 RGB to greyscale conversion

4.3.4 Results of Character Segmentation Component

DWT decomposes signal into different segments in frequency domain, while in 2D DWT, input images decomposed into four modules or sub-bands. The average segments (LL) and LL, HL, HH (detail segments) are demonstrated in Fig. 4.7.

The images decomposed using DWT is then subjected to Hough transforms wherein the lines are detected in the image. When the line is detected, it marks a horizontal line to the image. The output images after the application of Hough transform with the lines are provided in Fig. 4.8.

Text-line segmentation methods are used to perform tasks such as RGB to greyscale, conversation of greyscale to binary and finally text region segmentation. The Otsu's thresholding method iterates through the values of threshold and is used to assess the spread measure for the levels of pixels in each threshold side.

After binarization and segmentation of characters and lines, a hybrid approach which combined discrete wavelet transform (DWT) and Hough transforms was used wherein DWT decomposed the signals into various components in the frequency domain, and the two-dimensional DWT further decomposes the image into four sub-bands. Firstly, the image is segmented in a row-wise fashion which means line segmentation and the characters are segmented through column-wise segmentation of each row. Following DWT, Hough transforms are applied which enables the generation of Hough image from the binarized DWT edge map image. Certain parameters such as the theta, peaks and rho are initialized.

According to Choudhary et al. [Cho, 13], character segmentation is deemed as one important step when it comes to the overall character identification process. Grafmüller and Beyerer [Gra, 13] stated that character segmentation is used in character identification in various fields such as electronics, automobiles and pharmaceuticals. The previous researchers also reveal that line and character segmentation is an important step as characters that are wrongly segmented could cause several errors in terms of classification. However, the previous researchers have assumed that their models work only when the text region is specified.

However, in the present research, it is revealed that the proposed method which combines DWT and Hough transforms has the capability to segment characters without even specifying the text region. Previous researchers have also utilized the hybrid approach of NP character segmentation which is proposed by Vishwanath et al. [Vis, 12]. In the previous approach, the preprocessing of images is followed by vertical and horizontal segmentation which segments the NP characters. Furthermore, the authors have utilized the Hough transforms for performing vertical and horizontal segmentation.

However, the previous researchers are constrained to specific images pertaining to number plates. However, the present research considers a set of images which are with various characteristics. Segmenting several images with different characteristics is often considered as a daunting task which is due to the presence of noise, background and so on. However, the present research is an attempt where an image of any kind with characters can be used for segmenting characters. The images that

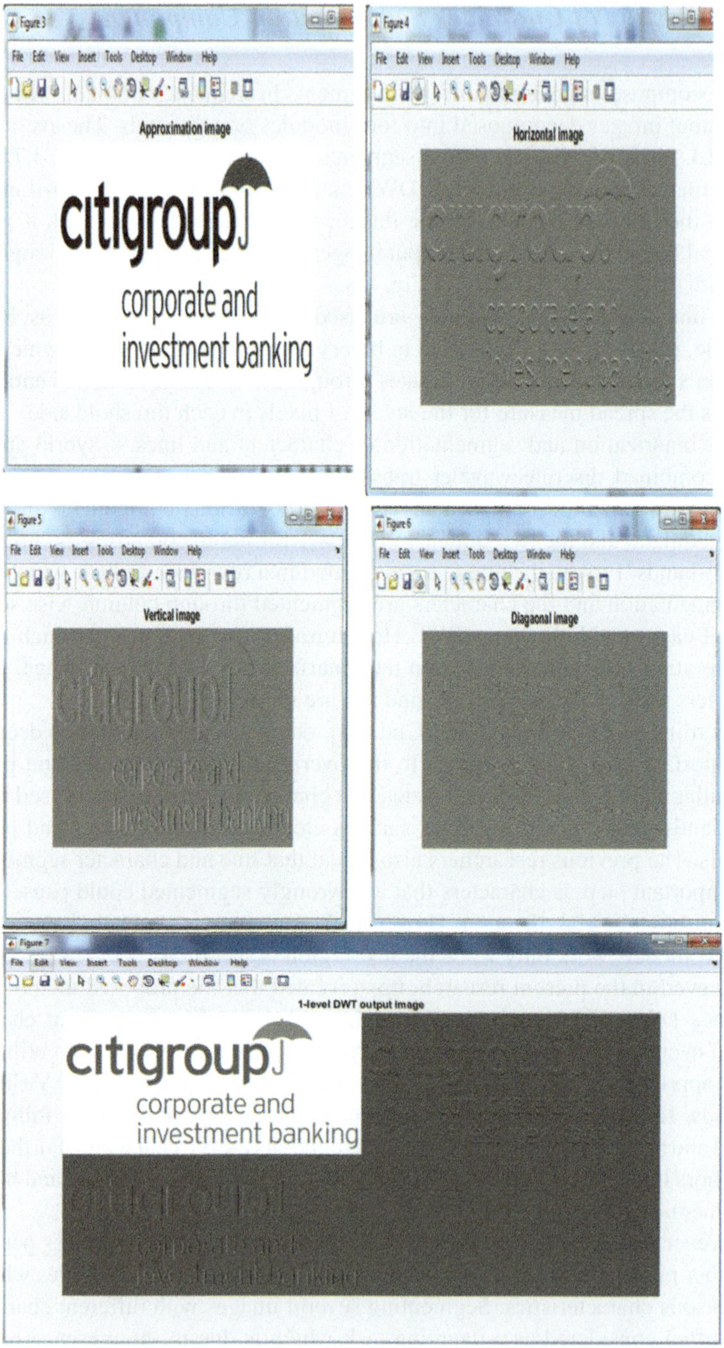

Fig. 4.7 Application of DWT

Fig. 4.8 Application of Hough transforms

are selected from the Dredze data set are based on the careful examination of features in the images wherein the proposed character segmentation approach is designed to segment characters from any image. The selection of different images with different characteristics is similar to research by Jung et al. [Jun, 04] who considered different images from newspapers, video frames, envelopes and so on. The previous research similar to the present research has also used different image types such as web images, scanned colour images and document images. Such usage of different images is to ensure the reliability of the proposed character segmentation method. However, the present research is a combination of various methods in a sequence such as character segmentation, recognition and shape-based feature extraction. The efficiency of the model is elaborated in Chap. 6 which will cover the overall performance and efficiency of the approach. The overall output for HAM and SPAM images are provided in Tables 4.1 and 4.2.

4.4 Summary

In the present section, the hybrid-based character segmentation approach is elaborated wherein the section covered the proposed approach which incorporated the RGB to greyscale conversion, binarization and removal of interconnected components, discrete wavelet transform (DWT), Hough-based line and character segmentation and information on the overall hybrid algorithm. With all the aforementioned concepts elaborated, the experimental set-up was elaborated with the experimental task, and the results of the character segmentation component are further elucidated.

Table 4.2 Results for SPAM images ($n = 5$)

SPAM							
Correct rate (CR)	Error rate	Sensitivity	Specificity	Precision	Recall	F-measure	Accuracy
82.261	17.72	100	79.1%	0.89	1	0.9524	95.161
82.321	17.62	100	78	0.902	1	0.958	95.6
74.2	25.81	100	68	0.91	1	0.95	95.2
82.25	17.75	66.6	86	0.88	0.67	0.75	95.13
80.64	19.35	91.67	78	0.863	0.916	0.9128	93.548

Table 4.1 Results for HAM images ($n = 5$)

HAM							
Correct rate (CR)	Error rate	Sensitivity	Specificity	Precision	Recall	F-measure	Accuracy
82.3	17.7	100	78%	0.909	1	0.95	96.7
85.4	14.5	100	82	0.909	1	0.9523	96.7742
82.26	17.742	83.33	82	0.9091	0.833	0.87	95.16
77.42	22.6	75	78	0.91	0.75	0.822	96.8
83.9	16.13	75	86	0.91	0.75	0.823	93.56

References

[Ots, 79] Otsu, N. (1979). A threshold selection method from gray-level histograms. *IEEE Transactions on Systems, Man, and Cybernetics, 9*(1), 62–66.

[Ran, 15] Rani, S., & Kaur, R. (2015). Review: Audio noise reduction using filters and discrete wavelet transformation. *Journal of the International Association of Advanced Technology and Science, 16*(6).

[Jef, 12] Jeffrey, Z., Ramalingam, S., & Bekooy, N. (2012). *Real-time DSP-based license plate character segmentation algorithm using 2D Haar wavelet transform.* Garden City, UK: Welwyn.

[Kas, 12] Kashyap, N., & Sinha, G. R. (2012). Image watermarking using 2-level DWT. *Advances in Computational Research, 4*(1), 42–45.

[Tu, 15] Tu, C. (2015). Enhanced Hough transforms for image processing. Paris-Est.

[Sah, 10] Saha, S., Basu, S., Nasipuri, M., & Basu, D. K. (2010). A Hough transform based technique for text segmentation. *Journal of Computing, 2*(2), 135–141. Retrieved from http://arxiv.org/abs/1002.4048.

[Fer, 15] Fernandez, J. A., Boddeti, V. N., Rodriguez, A., & Kumar, B. V. K. V. (2015). Zero-aliasing correlation filters for object recognition. *IEEE Transactions on Pattern Analysis and Machine Intelligence, 37*(8), 1702–1715. Retrieved from http://ieeexplore.ieee.org/document/6967788/.

[Kum, 02] Kumar, V. B. V. K., Savvides, M., Venkataramani, K., & Chunyan, X. (2002). Spatial frequency domain image processing for biometric recognition. In *Proceedings international conference on image processing. [Online]* (pp. I-53–I-56). New York: IEEE. Retrieved from http://ieeexplore.ieee.org/document/1037957/.

[Mah, 96] Mahalanobis, A., Vijaya Kumar, B. V. K., & Sims, S. R. F. (1996). Distance-classifier correlation filters for multiclass target recognition. *Applied Optics, 35*(17), 3127. Retrieved from https://www.osapublishing.org/abstract.cfm?URI=ao-35-17-3127.

[Kum, 06] Kumar, B. V. K. V., Savvides, M., & Chunyan, X. (2006). Correlation pattern recognition for face recognition. *Proceedings of the IEEE, 94*(11), 1963–1976. Retrieved from http://ieeexplore.ieee.org/document/4052477/.

[Tho, 07] Thornton, J., Savvides, M., & Kumar, B. V. K. V. (2007). A Bayesian approach to deformed pattern matching of Iris images. *IEEE Transactions on Pattern Analysis and Machine Intelligence, 29*(4), 596–606. Retrieved from http://ieeexplore.ieee.org/document/4107564/.

[Rod, 08] Rodriguez, M. D., Ahmed, J., & Shah, M. (2008). Action MACH a spatiotemporal maximum average correlation height filter for action recognition. In *IEEE Conference on Computer Vision and Pattern Recognition. [online]*. New York: IEEE. Retrieved from https://arxiv.org/pdf/1411.2316.pdf.

[Bod, 13] Boddeti, V. N., Kanade, T., & Kumar, B. V. K. V. (2013). Correlation filters for object alignment. In *2013 IEEE Conference on Computer Vision and Pattern Recognition. [online]* (pp. 2291–2298). New York: IEEE. Retrieved from http://ieeexplore.ieee.org/document/6619141/.

[Hen, 12] Henriques, J. F., Caseiro, R., Martins, P., & Batista, J. (2012). Exploiting the Circulant structure of tracking-by-detection with kernels. In *Computer Vision—ECCV 2012. [online]* (pp. 702–715). Berlin: Springer. Retrieved from http://link.springer.com/10.1007/978-3-642-33765-9_50.

[Bol, 10] Bolme, D., Beveridge, J. R., Draper, B. A., & Lui, Y. M. (2010). Visual object tracking using adaptive correlation filters. In *2010 IEEE Computer Society Conference on Computer Vision and Pattern Recognition. [online]* (pp. 2544–2550). New York: IEEE. Retrieved from http://ieeexplore.ieee.org/document/5539960/.

[Rod, 13] Rodriguez, A., Boddeti, V. N., Kumar, B. V. K. V., & Mahalanobis, A. (2013). Maximum margin correlation filter: A new approach for localization and classification. *IEEE Transactions on Image Processing, 22*(2), 631–643. Retrieved from http://ieeexplore.ieee.org/document/6310059/.

[Hen, 13] Henriques, J. F., Carreira, J., Caseiro, R., & Batista, J. (2013). Beyond hard negative mining: Efficient detector learning via block-Circulant decomposition. In *2013 IEEE International Conference on Computer Vision. [online]* (pp. 2760–2767). New York: IEEE. Retrieved from http://ieeexplore.ieee.org/document/6751454/.

[Bod, 14] Boddeti, V.N., & Kumar, B.V.K.V. (2014). *Maximum margin vector correlation filter*. [online]. Arxiv. Retrieved March 16, 2018, from https://arxiv.org/pdf/1411.2316.pdf.

[Rev, 13] Revaud, J., Douze, M., Schmid, C., & Jegou, H. (2013). Event retrieval in large video collections with Circulant temporal encoding. In *2013 IEEE Conference on Computer Vision and Pattern Recognition. [online]* (pp. 2459–2466). New York: IEEE. Retrieved from http://ieeexplore.ieee.org/document/6619162/.

[Can, 86] Canny, J. (1986). A computational approach to edge detection. *IEEE Transactions on Pattern Analysis and Machine Intelligence, 8*(6), 679–698. Retrieved from http://ieeexplore.ieee.org/lpdocs/epic03/wrapper.htm?arnumber=4767851.

[Dre, 07] Dredze, M., Gevaryahu, R., & Elias-Bachrach, A. (2007). Learning fast classifiers for image Spa. In *Proceedings of the conference on email and anti-spam*. New Delhi: CEAS.

[Cam, 09] De Campos, T., Babu, B. R., & Varma, M. (2009). Character recognition in natural images. *Proceedings of the International Conference of Computer Vision Theory and Applications, 3*(3), 2000–2003. Retrieved from http://eprints.pascal-network.org/archive/00005365/.

[Cho, 13] Choudhary, A., Rishi, R., & Ahlawat, S. (2013). A new character segmentation approach for off-line cursive handwritten words. *Procedia Computer Science, 17*, 88–95. Retrieved from http://linkinghub.elsevier.com/retrieve/pii/S1877050913001464.

[Gra, 13] Grafmüller, M., & Beyerer, J. (2013). Performance improvement of character recognition in industrial applications using prior knowledge for more reliable segmentation. *Expert Systems with Applications, 40*(17), 6955–6963. Retrieved from http://linkinghub.elsevier.com/retrieve/pii/S0957417413003849.

[Vis, 12] Vishwanath, N., Somasundaram, S., Baburajani, T. S., & Nallaperumal, N. K. (2012). A hybrid Indian license plate character segmentation algorithm for automatic license plate recognition system. In *2012 IEEE International Conference on Computational Intelligence and Computing Research. [online]* (pp. 1–4). New York: IEEE. Retrieved from http://ieeexplore.ieee.org/document/6510322/.

[Jun, 04] Jung, K., In Kim, K., & Jain, A. K. (2004). Text information extraction in images and video: a survey. *Pattern Recognition, 37*(5), 977–997. Retrieved from http://linkinghub.elsevier.com/retrieve/pii/S0031320303004175.

Chapter 5
Character Recognition

5.1 Introduction

In the present chapter, contour analysis with improved local binary pattern is proposed for text recognition and visual feature extraction. Based on the examination of literature article, the aforementioned techniques were selected and were hypothesized to improve the efficiency of feature extraction. In order to get the smooth contours of images, a double filter bank, the Laplacian Pyramid (LP), followed by Directional Filter Bank (DFB) provides better multiscale decomposition and removes the low frequency. The present chapter covers the following sections—Sect. 5.2 covers the proposed contour analysis with improved local binary pattern technique; Sect. 5.3 elaborates on the experimental results and analysis, and Sect. 5.4 covers the summary.

5.2 Proposed Method—Using a Combination of Text Recognition and Visual Feature Extraction for Character Recognition

The research proposed contour analysis with an improved local binary pattern for text recognition and visual feature extraction. To get the smooth contours of images, a double filter bank, the Laplacian Pyramid (LP), followed by Directional Filter Bank (DFB) provides better multiscale decomposition and removes the low frequency. LBP considers the effects of central pixels, present complete structure patterns to enhance the discriminative ability.

M. Rajalingam, *Text Segmentation and Recognition for Enhanced Image Spam Detection*, EAI/Springer Innovations in Communication and Computing,
https://doi.org/10.1007/978-3-030-53047-1_5

5.2.1 *Contour Analysis*

In the recognition phase, the contour analysis is used to recognize the characters after segmenting each character. Contour analysis includes the vector representation of each contour characters, and matching is done using the scalar product of each complex vector representation with a previously stored pattern in the training phase [Sou, 14].

The contour analysis (CA) permits explaining, saving, comparing and locating the characters submitted in the shape of the outside contours. It is presumed that the contour includes the adequate info on the character form. The contour is the border of characters, a population of points (pixels), and dividing character from a backdrop that is acquired from thinning. The measures entailed in the contour analysis are debated as under:

1. Thinning/Contour Extraction
2. Vector Representation
3. Average gradient magnitude of contour pixels
4. Gradient direction the variance of contour pixels
5. Number of contour pixels

In a contour analysis, the contour is encrypted by the series comprising intricate numbers. On a contour, the point that is known as a starting point is set. Consequently, the contour is scanned (is permissible—clockwise), and every vector of offset is marked by an intricate number, $a + ib$. In the equation, a is the point offset on the x-axis and b is the point offset on the y-axis. The offset is marked regarding the earlier point.

Thus, vector-contour Γ of length k can be designated as follows:

$$\Gamma = (\gamma_0, \gamma_1, \ldots, \gamma_{k-1})$$

The function over a contour as over a vector of intricate numbers owns astonishing mathematical elements, in comparison with other modes of coding. Fundamentally, intricate coding is almost two-dimensional coding where the contour is described as a population of the EVs submitted in the two-dimensional coordinates. But a disagreement betwixt the function of scalar product for vectors and for intricate numbers is diverse. This situation, in addition, provides preference to CA procedures.

5.2.2 *Improved Local Binary Pattern*

Local Binary Pattern (LBP), primarily planned for effective texture categorization, gives a plain and effectual method for design identification [Liu, 10a]. Two most significant elements of the LBP operator are its sturdiness against brightness alterations and its computational ease.

For every pixel, the LBP is ascertained. This is a value that explains the regional spatial framework. It is stated as an integer value.

5.2.2.1 Calculation of LBP

To estimate the LBP for every pixel, its greyscale value is compared to a number of pixels in the vicinity. These pixels are ascertained by the kind of design employed. A comparison betwixt the centre pixel and one of the encircling pixels provides either a 1 or a 0, denoting whether the centre pixel has a higher greyscale value. By comparing every encircling pixel, a bit-string can be produced. Thus, every exclusive spatial framework has its own distinguishing value. Mathematically, the subsequent equations from the initial LBP article [Oja, 2000] explains the computation of this value.

$$s(x) = \begin{cases} 1, x \geq 0 \\ 0, x < 0 \end{cases}$$

$$\text{value} = \sum_{p=0}^{p-1} s\left(g_{\text{p}} - g_{\text{c}}\right) 2^{p}$$

In these equations, g_{c} is the greyscale value of the centre pixel and g_{p} for $p = 0, \ldots, P - 1$ are the greyscale values of the surrounding pixels [Mei, 12]. The character contours are shown in Fig. 5.1.

Vector depiction of characters is illustrated in Fig. 5.2. Texture dispensations that disregard the real stations or spatial structure in the image are not enough for fine identification. Besides, we use sub-blocks to portray the spatial structure of an image. Whereas the input image is separated into $M \times N$ non-overlapping sub-blocks, and LBP histograms info of sub-blocks were combined as the element descriptor. The LBP histogram of every block could represent the peripheral alteration, flatness or sharpness of areas, the presence of particular points and so on. In our database,

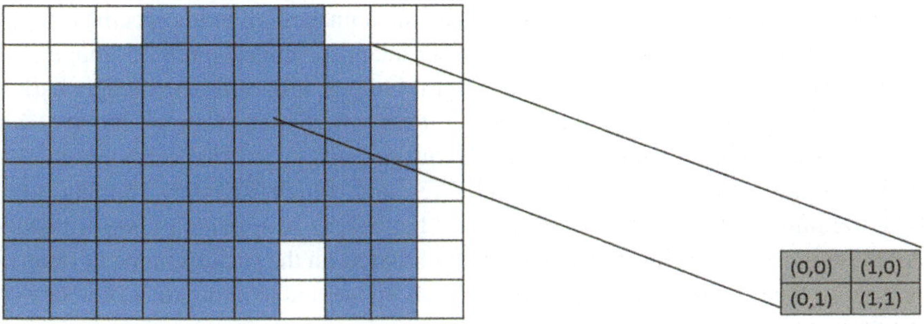

Fig. 5.1 Find LBP for each pixel

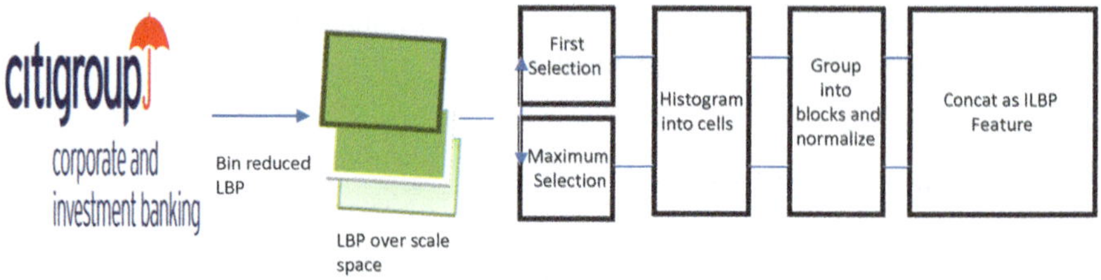

Fig. 5.2 Flow diagram of vector depiction of characters

every character from licence plate is standardized as a 64 × 64 pixels grey image following a sequence of pre-processing. Taking the low resolution of every character into account, we usually separate the 64 × 64 character image as 4 × 4, 8 × 8, or 6 × 6 sub-square blockings.

The modified LBP without rotation invariance and with concepts from HOG added. There are four main modified methods in our implementation: the creation of an LBP by averaging pixel intensity via integral image, reduction of the number of histogram bins, the creation of an LBP over scale space, and using a blocks and cells concept from HOG. In the bin reduction step, reduce the number of LBP histogram bins and add a magnitude map similar to that of HOG; thus, each pixel can express an integral magnitude and direction information, which we called edge type. The second step is to create a bin-reduced LBP over scale space, and the third step is to determine a meaningful scale in scale.

5.3 Experiment

5.3.1 Experimental Set-Up

For experimentation, 50 images were tested and were taken from the image spam data set [Dre, 07]. These images are in two groups that are 25 spam and 25 ham images. For the purpose of identifying both ham and spam images from the selected images of the Dredze data set, both ham and spam images are taken and are displayed in Fig. 3.3 for the sake of notifying which images fall in the specific categories. In the present research, all images were taken for training and 35 images for testing. The training images comprise several images with different image characteristics. This is in line with the testing images wherein each image has its own characteristics when it comes to noise, background and so on. Furthermore, the use of 35 samples for testing is in line with the research which used only 15 images for both training and testing, however, revealed better results in terms of performance. All these images are selected from the data set based on the various kinds of images present in the data set, the uniqueness of each image and to show the efficiency of the proposed ham spam detection method.

Fig. 5.3 Simulated results of character contours

5.3.2 Results of Thinning/Contour Extraction

The fundamental action is to make the boundary of characters 1 pixel thin. It observes the image pixel by pixel and erases the inner layers of black pixels on every character. The image is observed repeatedly until every character boundary is reduced to single pixel thickness. Simulated results of character contours are shown in Fig. 5.3.

5.3.3 Results of Vector Representation

In a contour analysis, the contour is described by the series comprising intricate numbers that are depicted by the vectors. On a contour, the point that is known as a starting point is set and then the contour is scanned (clockwise), and each vector of offset is observed by an intricate number. The offset is observed regarding the earlier point. Vector depiction of characters is illustrated in Fig. 5.4.

As per the tangible quality of characters, their contours are forever shut and cannot possess self-crossing. It permits describing a method of bypassing of a contour (to within a way—on or anticlockwise). The final vector of a contour forever guides to the beginning point. Each vector of a contour we will designate Elementary Vector (EV). And the series of intricate-valued numbers—Vector-Contour (VC). Simulated results of vector representation of characters are shown in Fig. 5.5.

5.3.4 Results of Average Gradient Magnitude of Contour Pixels

The initial conviction is that the average grade size of peripheral pixels is generally higher for text than non-text blocks. We can compute this mean value of each CC area with the formula that is furnished as follows:

$$V_{\text{avg}} = \frac{\sum_{(i,j)\in} V(i,j)}{m}, \left(\text{CL}(i,j) = \text{CL}_R\right)$$

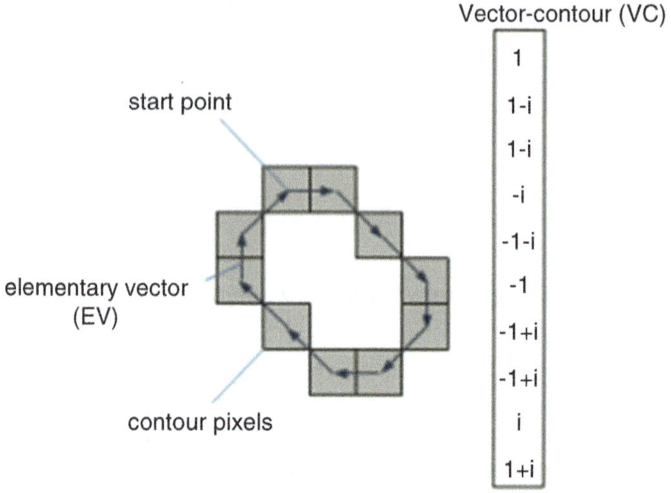

Fig. 5.4 Vector Representation of Characters

Fig. 5.5 Simulated results of vector representation of characters

where m is the number of pixels labelled with 1 CC class CL_R representing the region R. V_{avg} of a text region should be greater than $2*T$. Simulated results of average gradient of contour pixels are shown in Fig. 5.6.

Fig. 5.6 Simulated results of the average gradient of contour pixels

Fig. 5.7 Simulated results of Contour pixels

5.3.5 Results of Gradient Direction Variance of Contour Pixels

The text areas possess a higher grade way dispensation difference than graphic areas. The difference of an area can be calculated by the supreme and least peripheral grade way angle individually in a CC area. Within a text area, the difference must be higher than PI (180°), that is (max – min) > PI.

5.3.6 Results of Number of Contour Pixels

The third useful certitude we noted was that text block ought to possess extra peripheral pixels than certain non-text blocks. The peripheral pixel tallies in a text block, marked as the identical class within a CC area, must be higher than MAX (2 * W, 2 * H), where W and H are breadth and height of the CC area individually. Figure 5.7 shows simulated results of contour pixels.

Fig. 5.8 Simulated results of character recognition

5.3.7 Results of Character Recognition

The character recognition takes a text image as input and gives editable text document as output. The proposed character recognition primarily involves three steps: features extraction, features training, and feature matching. Here, two data sets are considered, one for training data set and another for test data set. The feature extraction is done in both cases. Features extracted from test data is compared with features extracted from training data to get the desired output. The output results of character recognition are shown in the Fig. 5.8.

5.4 Summary

In the present section, the hybrid-based character recognition approach is elaborated wherein the section covered the proposed approach which incorporated the contour analysis, character recognition along with the improved LBP (includes Thinning/Contour Extraction of Vector Representation, Average gradient magnitude of contour pixels, Gradient direction variance of contour pixels, Number of contour pixels,

Character recognition). With all the aforementioned concepts elaborated, the experimental set-up was elaborated with the experimental task, and the results of the character recognition component are further elucidated.

References

[Sou, 14] Soumya, K. R., Babu, A., & Therattil, L. (2014). License plate detection and character recognition using contour analysis. *International Journal of Advanced Trends in Computer Science and Engineering., 3*(1), 15–18.

[Liu, 10a] Liu, L., Zhang, H., Feng, A., Wan, X., & Guo, J. (2010a). Simplified local binary pattern descriptor for character recognition of vehicle license plate. In *2010 Seventh international conference on computer graphics, imaging and visualization. [online]* (pp. 157–161). New York: IEEE. Retrieved from http://ieeexplore.ieee.org/document/5576211/.

[Oja, 2000] Ojala, T., Pietikainen, M., & Maenpaa, T. (2000). Gray scale and rotation invariant texture classification with local binary patterns. In: Computer Vision—ECCV 2000. *Lecture Notes in Computer Science, 1842*, 404–420. Retrieved from http://link.springer.com/10.1007/3-540-45054-8_27.

[Mei, 12] Meijer, J. (2012). *License plate recognition using Local Binary Patterns.* [online]. Retrieved from https://esc.fnwi.uva.nl/thesis/centraal/files/f1317872649.pdf.

[Dre, 07] Dredze, M., Gevaryahu, R., & Elias-Bachrach, A. (2007). Learning fast classifiers for image Spa. In *Proceedings of the conference on email and anti-spam.* New Delhi: CEAS.

Chapter 6
Classification/Feature Extraction Using SVM and K-NN Classifier

6.1 Introduction

In this chapter the proposed visual feature extraction using improvised SVM and K-NN classifiers are discussed. Thus, the proposed is an automatic, stable, quick response automatic segmentation, followed by feature extraction and classification to detect spam from the images and the text. The K-NN classifier is used to extract features by predicting nearest neighbour, while SVM analyses the data for classification and regression. The present chapter covers the following sections—Sect. 6.2 covers the proposed improvised SVM and K-NN classifiers, Sect. 6.3 elaborates on the experimental results and Sect. 6.4 covers the summary.

6.2 Proposed Method: A Complete Character Segmentation Detection

6.2.1 Feature Extraction

The feature extraction is an integral part of the recognition system. The feature extraction process is to identify patterns by means of a minimum number of features that are effective in discriminating pattern classes. In our research the identification of characters by a hybrid machine learning (K-NN and SVM) approach is applied; each machine needs specific feature vectors. The K-NN is used to classify all data sets without training and then multiple class SVMs are performed only on the smaller data set with similar characters. The pictorial representation of feature extraction and classifier is shown in Fig. 6.1.

M. Rajalingam, *Text Segmentation and Recognition for Enhanced Image Spam Detection*, EAI/Springer Innovations in Communication and Computing,
https://doi.org/10.1007/978-3-030-53047-1_6

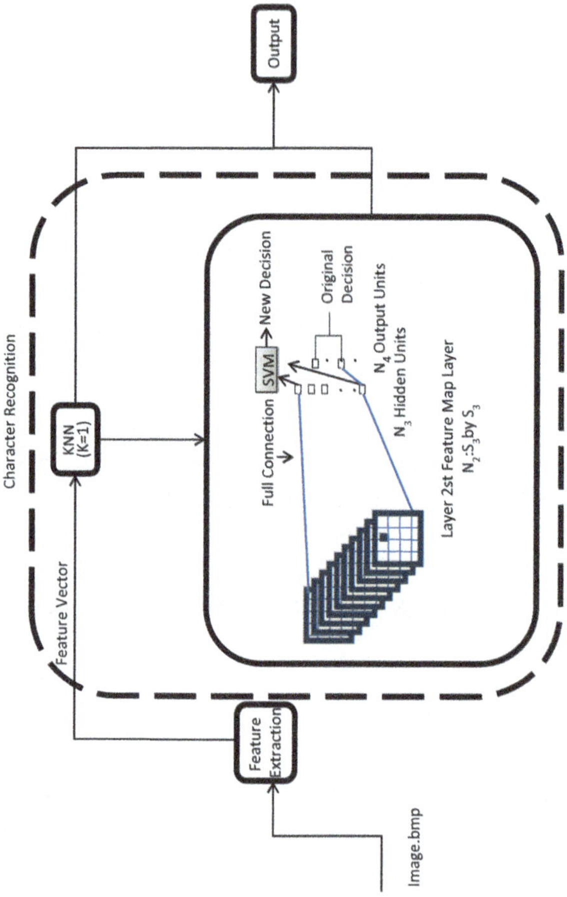

Fig. 6.1 Feature extraction and classifier technique

6.2.2 SVM

Support vector machines employ a non-linear mapping to alter the initial training data into a greater extent. It seeks for the linear maximum dividing hyperplane with a suitable number of support vectors (SVs). Info from two classes can forever be divided by a hyperplane. The SVM discovers this hyperplane employing support vectors and edges. Although the training time of even the quickest SVMs can be very high; they are extremely precise because of their skill to sample intricate non-linear decision borders. An SVM is fundamentally a binary classifier with the separate operation being the weighted mixture of kernel operations overall training models. The weights (coefficients) are studied by quadratic programming with the objective of magnifying the margin in feature space. Following the study, the models of non-zero weights are known as support vectors (SVs) that are saved and employed in categorization. Besides, we have employed linear kernel operation. Here, the kernels change the initial information into a greater extent feature space. Even if the initial information are non-linear, the changed information is divisible by a hyperplane in feature space.

With this idea as the foundation, support vector machines have confirmed to attain fine generality functioning with no previous knowledge of the information. The rule of an SVM is to map the input stats onto a greater extent feature space non-linearly connected to the input space and establish a dividing hyperplane with the supreme margin between the two classes in the feature space.

6.2.3 Nearest Neighbour Search

The K-Nearest Neighbour's algorithm is a procedure for categorizing objects grounded on the nearest training in the feature space. Also, it is a kind of case-grounded studying by connecting unfamiliar design to the familiar based on certain distance. It is used to calculate a categorization operation by investigating the marked training points as nodes or anchor points in the *n*-dimensional space, where *n* is the feature dimension. Furthermore, the Euclidean distance between the recollect point and the pixel point is estimated and then k-closest neighbours are found. Later, the attained distances are ranked in ascending order, and the mentioning points matching with the k-smallest Euclidean distances are taken. The K-NN categorization separates information into a training data set and test data set.

6.3 Experiment

6.3.1 Experimental Set-Up

For experimentation, 75 images were tested and were taken from the image spam data set [Dre, 07]. These images are in two groups that are 40 spam and 35 ham images. For the purpose of identifying both ham and spam images from the selected images of the Dredze data set, both ham and spam images. In the present research, all images were taken for training and 35 images for testing. The training images comprise several images with different image characteristics. This is in line with the testing images wherein each image has its own characteristics when it comes to noise, background and so on. Furthermore, the use of 35 samples for testing is in line with the research which used only 20 images for both training and testing, however, revealed better results in terms of performance. All these images are selected from the data set based on the various kinds of images present in the data set, the uniqueness of each image and to show the efficiency of the proposed ham spam detection method.

6.3.2 Results of K-NN and SVM Classifier

The k-nearest neighbour classifier has been employed here to categorize among various classes or sets of characters where the classifier is primarily trained by employing certain training data sets. For this objective, certain images with identified character are selected for training, and their arrays are ascertained employing the above procedure of value assessment explained in part 6.2. The K-NN classifier gives the optimal when the number of neurons at the hidden layer is 20.

A summary of classification accuracies obtained for three data sets are discussed. It shows that Multilayer Perceptron (MLP) yields the highest performance in most cases. It gives a recognition rate of 96.7%. In MLP, a number of neurons at the hidden layer is chosen experimentally for every data set. The selection of feature sets, feature optimization, post-processing and/or pre-processing can significantly contribute to the classification accuracy for all classifiers. Figure 6.2 shows the results of feature extraction.

Figure 6.3 shows the simulated results of proposed classifier technique. The text extraction and character segmentation have been carried out by both groups like training and testing. The *K*-NN ($K = 1$) with Euclidean distance functions was applied in the classification phase, and the multiclass SVMs were conducted.

K-NN classifier is used to verify whether vectors corresponding to the upper- and lower-case versions of the same letter are distributed in neighbouring regions of the feature space or not. The more the two versions of the letter are similar in shape, the more their vectors are overlapping and able to join in a single class.

Fig. 6.2 Simulated results of feature extraction

6.4 Summary

In the present section, the hybrid-based visual feature extraction and classification is elaborated wherein the section covered the proposed approach which incorporated the use of improvised SVM and K-NN classifier. Moreover, the process identified patterns via feature extraction method by means of a minimum number of features that are effective in discriminating pattern classes. With all the aforementioned concepts elaborated, the experimental set-up was elaborated with the experimental task, and the results of the character recognition component are further elucidated.

Fig. 6.3 Simulated results of classifier technique

Reference

[Dre, 07] Dredze, M., Gevaryahu, R., & Elias-Bachrach, A. (2007). Learning fast classifiers for image Spa. In *Proceedings of the conference on email and anti-spam*. New Delhi: CEAS.

Chapter 7
Experimentation and Result Discussion

7.1 Introduction

The image segmentation, recognition and classification are a complex process that may be affected by many factors. This chapter discusses the obtained results of proposed framework (CSRC) with performance evaluation metric. The emphasis is placed on the support vector machine and K-nearest neighbour's classification approach which depicts how this technique is used for improving classification accuracy.

7.2 Evaluation

In this chapter, the results of performance evaluation for solving the issues of spam and ham email detection are discussed.

The benchmarks that were employed to evaluate the functioning of this work are accuracy (AC), recall (R), precision (P), false positive (FP), false negative (FN), true positive, true negative, correct rate and error rate (ERR). The term TP that gauges the count of spam messages is rightly categorized as spam. TN gauges the number of non-spam emails that is rightly categorized as non-spam. FP gauges the number of miscategorized non-spam emails. FN gauges the number of miscategorized spam emails. Accuracy gauges the percentage of rightly recognized spam and non-spam messages. F-measure gauges the weighted average of precision and recall. These parameters were estimated by employing Eqs. (7.1–7.4), individually. Therefore, the output of the five different SPAM images after the segmentation process is tabulated in Table 7.1.

M. Rajalingam, *Text Segmentation and Recognition for Enhanced Image Spam Detection*, EAI/Springer Innovations in Communication and Computing,
https://doi.org/10.1007/978-3-030-53047-1_7

Table 7.1 Description of expressions for precision and recall

		Classification	
		SPAM email	HAM email
Obtained result	SPAM email	True positive	False positive
	HAM email	False negative	True negative

$$\text{Precision} = \frac{\text{TP}}{\text{TP} + \text{FP}} \tag{7.1}$$

$$\text{Recall} = \frac{\text{TP}}{\text{TP} + \text{FN}} \tag{7.2}$$

The F-measure is evaluated using precision and recall value, and correspondingly termed as balanced f-score or traditional F-measure.

$$\text{F} - \text{measure} = 2 * \frac{\text{Precision} * \text{Recall}}{\text{Precision} + \text{Recall}} \tag{7.3}$$

$$\text{Accuracy} = \frac{\text{TN} + \text{TP}}{\text{TP} + \text{FP} + \text{FN} + \text{TN}} \tag{7.4}$$

Eq. (7.4) depicts how well a binary classification test correctly, i.e. what percentage of predictions that is correct. The F-measure, precision, accuracy and recall values are calculated using the above-mentioned formula.

The error rate is the percentage of examples that the model has misclassified divided by the total number of examples:

$$\text{Error rate} = \frac{\text{FP} + \text{FN}}{\text{TP} + \text{FP} + \text{FN} + \text{TN}} \tag{7.5}$$

In other words the amount of emails misclassified regardless of the mistakes made by the model [Her, 06].

The correct rate is the ratio of correctly classified samples to the classified samples, and the error rate is the ratio of incorrectly classified samples to the classified samples [Sre, 14].

$$\text{Correct rate} = \frac{\text{Correctly classified samples}}{\text{Classified samples}} \tag{7.6}$$

$$\text{Error rate} = \frac{\text{Incorrectly classified samples}}{\text{Classified samples}} \tag{7.7}$$

The F-measure, precision, accuracy, recall, correct rate and error rate values are calculated using the above-mentioned formula.

7.3 Experimentation

The information that is accessible in the spam and HAM-base data set is in both numeric and string arrangement. The sixty qualities in the data set depict proportionate frequencies of different prominent words and characters in emails. We wish to change these to Boolean values for the experiment. The quality will take a value 1 if the word or character is there in the email and 0 if it is not there in the email. To do this, we use a numeric to binary filter that will change all the numeric values to the binary. The changed data set is employed to train the classifier to discern spam from normal email by verifying the number of occasions of every word for all the spam and non-spam emails.

The Precision, Recall, Accuracy and F-measure are estimated for each of the respective classifiers. From the outcomes measured, the classifiers grounded on the percentage of rightly categorized cases.

7.4 Results Discussions

The recommended algorithm is tried with 150 images. For this, email images with text in it along with images taken from image spam data set have employed.

A set of various characters of different font type, font style, font size, noise in background image, with low resolution, occluded images, special characters and special symbol are taken for experimentation. As there are 150 images, 410 lines and 5280 characters involved, not all of them can be listed. Only a few images are shown as output of character segmentation. The number of characters in each image is not the same after segmentation.

The research presents the novel framework which is a combination of character segmentation, recognition and classification to solve the spam detection from the image. First, we have extracted the text character from image via DWT and skew detection. Specifically, the character segmentation from images is done using DWT which includes morphological dilation operators and the logical AND operators to remove the non-text regions. Furthermore, having reduced the size of images, skew detection especially applying a fusion of Hough transform with spatial frequency cross-correlation was proposed. Previously skew detection algorithms such as Hough transforms, clustering, projection profiles, wavelet decompositions, morphology, moments, space parallelograms and Fourier analysis work on the assumption that images are black and white and optimized for documents in which text is predominant and arranged in parallel straight lines. However, previous algorithms did not provide perfect solution but only if they were used on appropriate documents. For skew detection, especially Hough transform with spatial frequency cross-correlation, was proposed. The fusion-based proposed method considers both structure and texture in the image, rather than threshold, as a basis for dividing an image into connected regions or polygons. This segmentation process is used to isolate features of a specific shape within an image and detects the regular curves like circles, lines, ellipses and so on.

Second, we have recognized the character via text recognition and visual feature extraction approach which relies on contour analysis with improved LBP. It attempts to extract embedded text together with the visual features like colour, texture and shapes and hence used to calculate a similarity measure with a query image. The extracted features are used to train a classifier which would work online in labelling an incoming message as spam or legitimate. The suggested method is robustness in contrast to illumination variations as well as its computational simplicity. Moreover, the determined LBP of each pixel value describes the local spatial structure, and it is expressed as an integer value. LBP considers the effects of central pixels, present complete structure patterns towards enhancing the discriminative ability.

Third, the extracted features are classified using SVM with KNN classifier. Wherein KNN is used to extract features by predicting nearest neighbour, while SVM analyses the data for classification and regression. The KNN and SVM classifier is processed as a chain code, and the output is the identified characters and their associated log-likelihoods. In our study we have used different formats of input images such as .jpg, .png and .bmp. Initially we processed the image with the help of binarization and thinning operation. This thinned image is undertaking a feature extraction process of chain coding. The chain-coded image kept in a file is then passed through a classifier. The document gets segmented as lines, whereas every line gets segmented into distinct characters. Moreover, the segmented files are scanned and line exists in image files is extracted and fed as an input to character segmentation. The segmentation of characters is done individually in every line. Moreover, the extracted characters that are needed to be identified are fed as input to the character recognizing sector.

Further the recognized character is given as input to the classification approach. Thus the proposed method is an automatic, stable, quick response automatic segmentation, followed by feature extraction and classification to detect spam from the images and the text. The proposed method was compared with other traditional methods.

The performance evaluation parameters help to choose a better technique. Along with the evaluation parameters, the simulation error helps to decide better technique. The obtained results are evaluated based on the performance metrics such as sensitivity, specificity, precision, recall, F-measure and accuracy. The values of correct rate (CR) and error rate (ER) are also evaluated for these proposed techniques. It is observed that the errors are on the minimum range for these proposed segmentation techniques. Table 7.2 shows that the obtained results are evaluated based on the performance metrics such as sensitivity, specificity, precision, recall, F-measure and accuracy.

The average value of performance metrics obtained for this proposed algorithm is about 82.2 CR, 17.73 ER, 86.6% sensitivity, 81% specificity, 0.909 precision values, 0.866 recall, F-measure of about 0.883 and accuracy of about 95.79%. The model improves output quality in terms of both sensitivity and specificity.

Table 7.2 Results for SPAM images ($n = 5$)

SPAM							
Correct rate (CR)	Error rate	Sensitivity	Specificity	Precision	Recall	F-measure	Accuracy
82.3	17.7	100	78	0.909	1	0.95	96.7
85.4	14.5	100	82	0.909	1	0.9523	96.7742
82.26	17.742	83.33	82	0.9091	0.833	0.87	95.16
77.42	22.6	75	78	0.91	0.75	0.822	96.8
83.9	16.13	75	86	0.91	0.75	0.823	93.56

Table 7.3 Results for HAM images ($n = 5$)

HAM							
Correct rate (CR)	Error rate	Sensitivity	Specificity	Precision	Recall	F-measure	Accuracy
82.261	17.72	100	79.1	0.89	1	0.9524	95.161
82.321	17.62	100	78	0.902	1	0.958	95.6
74.2	25.81	100	68	0.91	1	0.95	95.2
82.25	17.75	66.6	86	0.88	0.67	0.75	95.13
80.64	19.35	91.67	78	0.863	0.916	0.9128	93.548

Similarly, the output of the five different HAM images after the segmentation process is tabulated in Table 7.3 which has the average value obtained with these proposed algorithms of 82.2 CR, 17.73 ER, 86.6% sensitivity, 81% specificity, 0.909 precision, 0.866 recall, F-measure of about 0.883 and accuracy of about 95.79%.

An algorithm which has a highest accuracy will be considered as the enhanced approach with better classification capability. Therefore, Table 7.4 compared the accuracy range of few existing classifier studies. Though ANN, SVM, decision classifier method developed by Zhang et al. [Zha, 14b] have maximum accuracy of 94% than other classifiers, it consumes more time to build model. Moreover, Naive Bayes framed by Rusland et al. [Rus, 17] has least an accuracy of about 82.88% than other classifiers. Hence, our proposed method achieves the accuracy rate of about 95.79% and consumes less time from which it is concluded that it is the best classifier concerning accuracy.

The performance of proposed method has been compared with traditional method and illustrated in Fig. 7.1. The conventional techniques such as the Naïve Bayes, AD tree, SMO, Random Tree, ANN, SVM and Decision tree have the accuracy rate of about 82.88%, 91.60%, 92.63%, 91.54%, 94.38%, 94.42%, and 94.27%, respectively. However, our proposed technique achieved the enhanced accuracy rate of about 95.79% which represents that it outperformed the conventional methods.

Overall accuracy of the proposed method is shown in Fig. 7.2. The screenshot was illustrated by calculating performance measure.

The performance measure of sensitivity, specificity and accuracy of the spam images for the input data sets 1, 2, 3, 4 and 5 is demonstrated in Fig. 7.3. From the graph analysis, it is clear that the first input data set achieved the sensitivity of 100%, specificity of about 78% and accuracy range of about 96.7%. Similarly, the sensitivity,

Table 7.4 Comparison with the existing approach

Author	Method	Accuracy
Zhang et al. [Zha, 14a]	ANN, SVM, decision tree	94.38, 94.42, 94.27
Wu [Wu, 09]	ADTree	91.60
Lekha and Prakasam [Lek, 16]	SMO	92.63
Sharma and Arora [Sha, 13]	RANDOMTREE	91.54
Rusland et al. [Rus, 17]	naïve Bayes classifier	82.88%
[Author 19]	Proposed method	95.79884

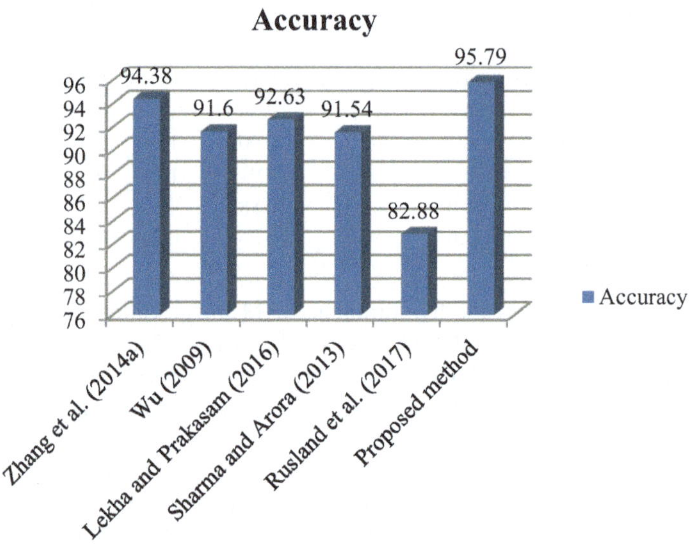

Fig. 7.1 Performance comparison with the existing method

specificity and accuracy of the second data set are 100%, 82% and 96.77%, respectively, and for the third data set, they are 83.33%, 82% and 95.16%, respectively. Moreover, for the fourth and fifth data sets, sensitivity, specificity and accuracy are in the range of 75%, 78% and 96.80% and 75%, 86% and 93.56%, respectively.

Similarly, the performance measures CR and ER for the five input data sets of the spam images are demonstrated in Fig. 7.4. From this graph, the CR and ER for the first data set are about 82.3 and 17.7 and for second and third input data sets, CR and ER are 85.4, 14.5 and 82.26, 17.7442, respectively. The fourth and fifth data sets have the CR and ER rate of about 77.42, 22.6 and 83.9, 16.13, respectively.

Moreover, Fig. 7.5 demonstrated the performance measure of precision, recall and F-measure of the SPAM images for the input data set. From this graph, it is shown that the F-measure value of the five input data sets is about 0.95, 0.9523, 0.87, 0.822 and 0.823, respectively. Recall value for the five input data sets is 1, 1, 0.833, 0.75 and 0.75, respectively. Moreover, the precision value of the five input data sets is about 0.909, 0.909, 0.9091, 0.91 and 0.91, respectively.

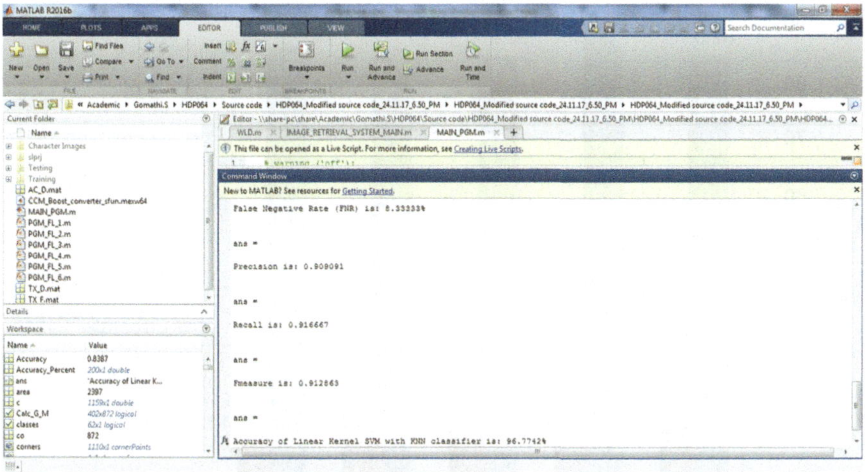

Fig. 7.2 Screenshot result of the proposed method

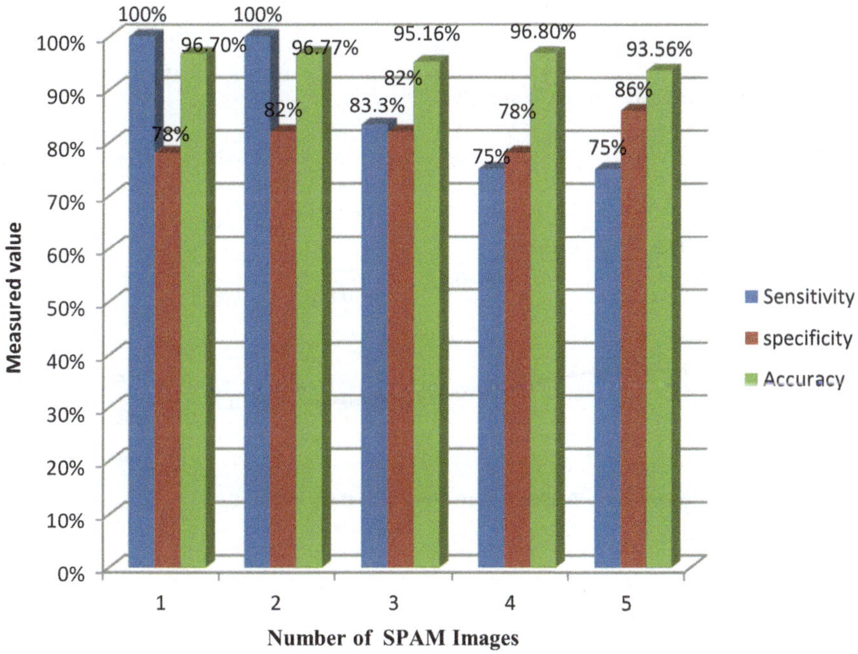

Fig. 7.3 Performance measure of sensitivity, specificity and accuracy

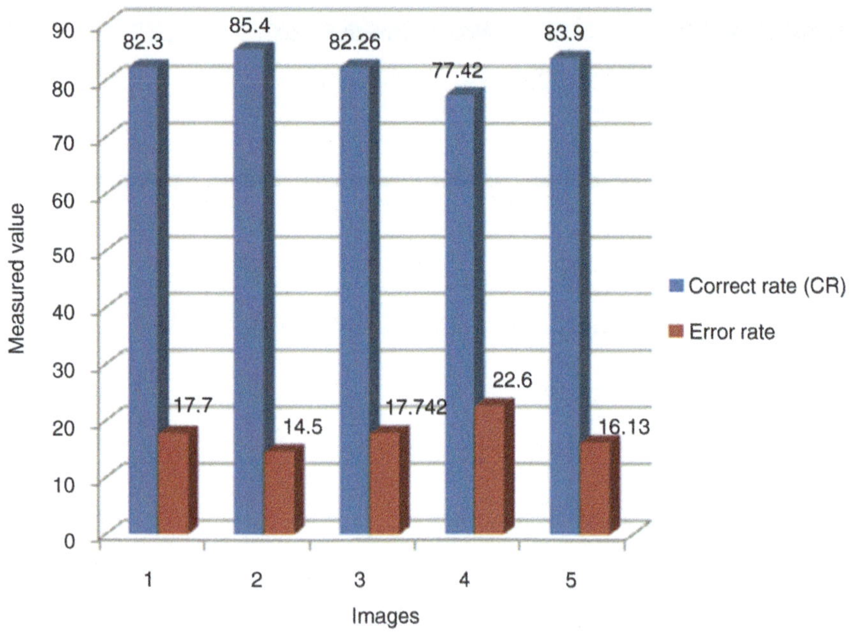

Fig. 7.4 Performance measure of CR and ER

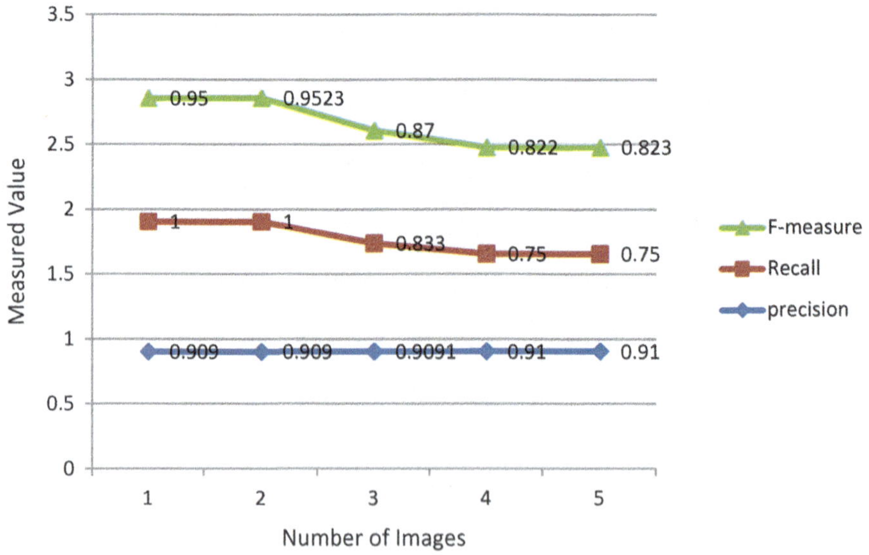

Fig. 7.5 Performance measure of precision, recall and F-measure

7.4.1 HAM Images

Figure 7.6 demonstrated the performance measure of CR and ER of the HAM images for the five input data sets. From this graph, the CR and ER for the first data set is about 82.26 and 17.72 and for second and third input data sets, CR and ER are 82.32, 17.62 and 74.21, 25.81, respectively. The fourth and fifth data sets have the CR and ER rate of about 82.25, 17.75 and 80.64, 19.35, respectively.

The performance measure of sensitivity, specificity and accuracy of the ham images for the input data sets 1, 2, 3, 4 and 5 are demonstrated in Fig. 7.7. From the graph analysis, it is clear that the first input data set achieved the sensitivity of 100%, specificity of about 79.1% and accuracy range of about 95.16%. Similarly, the sensitivity, specificity and accuracy for the second data set is 100%, 78% and 95.6%, respectively, and for the third data set, sensitivity, specificity and accuracy are in the range of 100%, 68% and 95.2%. Moreover, for the fourth and fifth data sets, sensitivity, specificity and accuracy range are 66%, 86%, 95.13% and 91.67%, 78% and 93.54%, respectively.

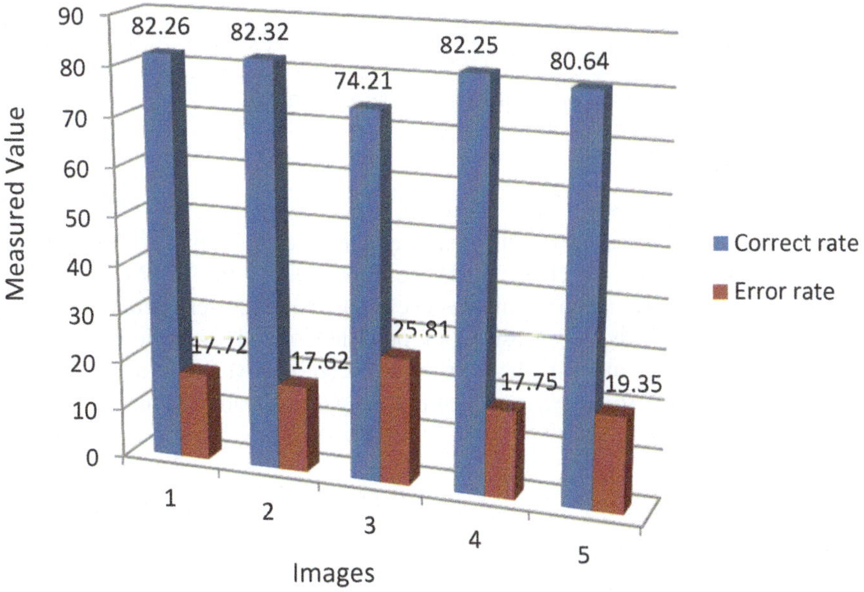

Fig. 7.6 Performance measure of CR and ER

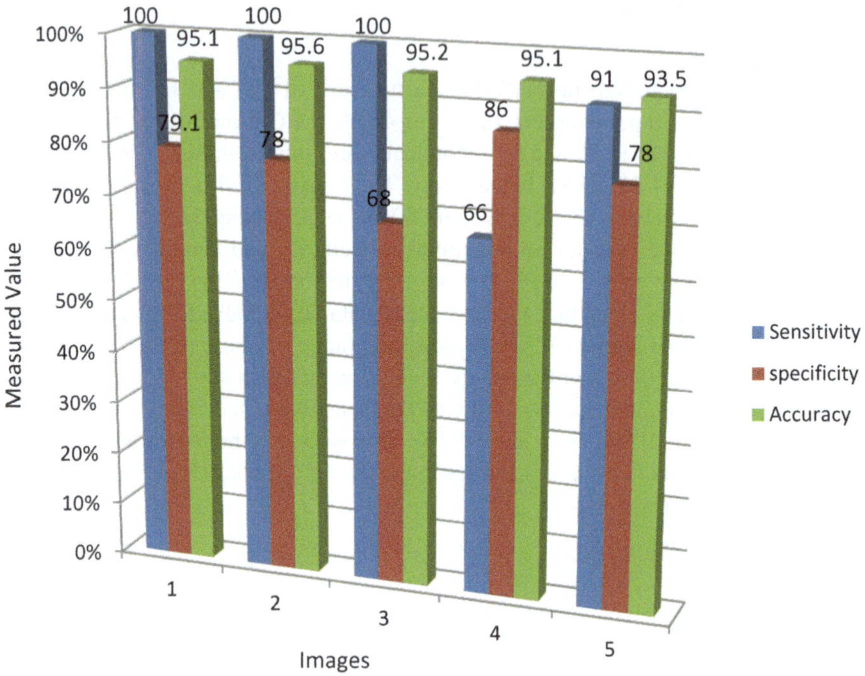

Fig. 7.7 Performance measure of sensitivity, specificity and accuracy

7.5 Summary

This chapter enlightened the results of segmentation, recognition and classification techniques are applied for character detection to identify the SPAM email. The accuracy of the classification techniques is evaluated by receiver operating characteristics which include the measures of precision, recall, f-measure, CR, ER, sensitivity and specificity. Finally, it provides a comparative study of how all the techniques discussed so far are used for improving classification accuracy.

References

[Her, 06] Hershkop, S. (2006). *Behavior-based email analysis with application to spam detection. [Online]*. Columbia: Columbia University. Retrieved from http://www1.cs.columbia.edu/~sh553/publications/final-thesis.pdf.

[Sre, 14] Sree Sharmila, T. (2014). *Efficient analysis of satellite image denoising and resolution enhancement for improving classification accuracy*. [Online]. Retrieved from http://shodhganga.inflibnet.ac.in/handle/10603/24460/.

[Zha, 14a]Zhang, Y., Wang, S., Phillips, P., & Ji, G. (2014a). Binary PSO with mutation operator for feature selection using decision tree applied to spam detection. *Knowledge-Based Systems, 64*, 22–31. https://doi.org/10.1016/j.knosys.2014.03.015.

[Wu, 09] Wu, C. H. (2009). Behavior-based spam detection using a hybrid method of rule-based techniques and neural networks. *Expert Systems with Applications, 36*(3 Part 1), 4321–4330. https://doi.org/10.1016/j.eswa.2008.03.002.

[Lek, 16] Lekha, K. C., & Prakasam, S. (2016). Prediction of Respondants' knowledge towards cyber security measures using various classification algorithms. *International Journal of IT and Knowledge Management, 10*(1), 1–5. Retrieved from http://csjournals.com/IJITKM/PDF10-1/1.Chitra.PDF.

[Sha, 13] Sharma, S., & Arora, A. (2013). Adaptive approach for spam detection. *IJCSI International Journal of Computer Science, 10*, 4. Retrieved from https://pdfs.semanticscholar.org/956c/dfa8574d01f0cdb2eaa5383ea5028a1eadc6.pdf.

[Rus, 17] Rusland, N. F., Wahid, N., Kasim, S., & Hafit, H. (2017). Analysis of Naïve Bayes algorithm for email spam filtering across multiple datasets. *IOP Conference Series: Materials Science and Engineering, 226*, 12091. Retrieved from http://stacks.iop.org/1757-899X/226/i=1/a=012091?key=crossref.262a040d608eaab52b8501086da85f26.

[Zha, 14b] Zhang, Y., Wang, S., Phillips, P., & Ji, G. (2014b). Binary PSO with mutation operator for feature selection using decision tree applied to spam detection. *Knowledge-Based Systems, 64*, 22–31. Retrieved from http://linkinghub.elsevier.com/retrieve/pii/S095070511400104X.

Chapter 8
Conclusion

The present chapter concludes the book with the findings and contributions of the current work briefly summarized, and the limitation of the proposed work is explained and possible future works are mentioned. The present research titled 'Integrated Methodology for Text Segmentation and Recognition for Enhanced Image Spam Detection' is based on the integration of several techniques so as to devise an application that would provide better results of email-based image spam detection. In this regard, the following summary of the research provides in-depth insights about the entire research.

With the increasing significance of internet around the word, email is one of the foremost approaches of communication among people. But due to the flood of online data, most of the people's inbox space is overwhelmed by unsolicited commercial email or spam email. The spam emails not only waste computing resources and network bandwidth of the internet users, but it also affects the larger scale, interrupt enterprises of standard system process. In order to detect image spam email, a novel framework is proposed which is a combination of character segmentation, recognition and classification technique (CSRC). The proposed framework exploits to take an advantage of processing low-level features and extraction of embedded text data. First, the text character is extracted from image by segmentation process which includes a combination of DWT and skew detection. Furthermore, logical AND as well as morphological dilation operators is applied towards removing the non-text regions. The size of input image is reduced by applying a fusion of Hough transform along with a spatial frequency cross-correlation approach, whereas the fusion-based approach deliberates both texture and structure of the input image. This segmentation process is utilized towards isolating the image features of a specific shape and detects the regular curves like circles, lines, ellipses and so on. Second, the character is recognized via text recognition and visual feature extraction approach which relies on contour analysis with improved LBP. This approach is robustness in contrast to illumination variations and its computational simplicity. Moreover, the local spatial structure is described

M. Rajalingam, *Text Segmentation and Recognition for Enhanced Image Spam Detection*, EAI/Springer Innovations in Communication and Computing,
https://doi.org/10.1007/978-3-030-53047-1_8

and LBP of each pixel value is determined. Third, the extracted text features are classified using KNN and SVM classifiers. Here, the KNN is utilized towards extracting the text features through predicting nearest neighbour, whereas SVM is used to analyse the text data for both classification and regression process.

In this research work, the experiment has been done based on the Dredze data set. This data set contains 3299 spam images and 2021 ham images. However, in processing stage we have eliminated the image which does not provide the enough information like image with no texture information and image size of less than 10 bytes. As a result, we have considered 2173 spam images and 1248 ham images for testing and validating the proposed CSRC framework. In addition, we have considered different formats of input images such as .jpg, .png and .bmp. The image is pre-processed via binarization and thinning operation. This thinned image is undertaking a feature extraction process of chain coding. The chain-coded image kept in a file is then passed through a classifier. The document is then segmented into lines and each line into individual characters. The documented is scanned and a line in the image file is extracted. The extracted line is given as input to character segmentation. Within each line the characters are segmented one by one. The extracted character that is still to be recognized is given as input to the Character Recognizing Module. Further the recognized character is given as input to the classification approach. Thus the proposed method is an automatic, stable, quick response automatic segmentation, followed by feature extraction and classification to detect spam from the images and the text.

The proposed methods are implemented using MATLAB programming language (version R 2016b), and the experimentations are performed on an Intel(R) Core (TM) i5 machine with a speed of 2.60 GHz and 8.0 GB RAM using Windows 8.1 64-bit Operating System. The performance of the proposed method is validated by the measure of different metrics such as sensitivity, specificity, precision, recall, F-measure, accuracy, error rate and correct rate. By the simulation, the obtained average value of the proposed framework is 86.6, 81, 90.9, 86.6, 88.3, 95.79, 17.73 and 82.2 (measured in %), respectively. Along with the evaluation parameters, the simulation error rate helps to decide better technique. The measurements are taken for the above-mentioned data set. Furthermore, the performance of the proposed method is compared with the existing approach which is based on the results reported in literature such as ANN, SVM, decision tree, ADTree, SMO, RANDOMTREE and Naıve Bayes classifier. By the literature, the measured accuracy of ANN, SVM, decision tree, ADTree, SMO, RANDOMTREE and Naıve Bayes classifier is 94.38, 94.42, 94.27, 91.60, 92.63, 91.54, and 82.88, respectively. The compared results indicate that the predictive performance for spam email detection using proposed framework (CSRC) outperforms than the existing method. Moreover, this proposed framework (CSRC) is flexible towards the detection of spam and ham email based on the image processing technique and higher in classifying spam from different categories of images. Additionally, the proposed method run significantly faster than the existing methods because the input to the spam data set includes seed pixels which considerably decrease the space of possible classification process, and it is able to reduce the impact of spam email and effectively filter out image spam messages.

Appendixes

```
% warning ('off');
clc;
clearall;
closeall;

%% Read the input image %
[filename,pathname]=uigetfile( {'*.png'; '*.bmp';'*.tif';'*.jpg'});
Input_Image=imread([pathname,filename]);
Fig. (1),imshow(Input_Image,[]);  %%%% show the image  %%%%%
title('original image');
[M N O]=size(Input_Image);
Input_To_Gray = rgb2gray(Input_Image);
%% RGB TO GRAY CONVERSION %%%%
if O==3
   G=rgb2gray(Input_Image);
else
   G=Input_Image;
end
Fig. (2),imshow(G,[]);  %%%% show the image  %%%%%
title('grayscale image');

%% DWT %%%%
[ll1,hl1,lh1,hh1]=dwt2(G,'haar');
Fig. (3),imshow(ll1,[]);
title('Approximation image');
Fig. (4),imshow(hl1,[]);
title('Horizontal image');
Fig. (5),imshow(lh1,[]);
title('Vertical image');
Fig. (6),imshow(hh1,[]);
title('Diagaonal image');
```


M. Rajalingam, *Text Segmentation and Recognition for Enhanced Image Spam Detection*, EAI/Springer Innovations in Communication and Computing,
https://doi.org/10.1007/978-3-030-53047-1

```
d1=[ll1,hl1;lh1,hh1];
Fig. (7),imshow(d1,[]);
title('1-level DWT output image');

%% Skew detection %%%%
[H,T,R] = hough(G);
Fig. (8),imshow(G,[]);
P  =houghpeaks(H,5,'threshold',ceil(0.3*max(H(:))));
lines = houghlines(G,T,R,P,'FillGap',5,'MinLength',7);
Fig. (9), imshow(G), hold on
max_len = 0;
for k = 1:length(lines)
xy = [lines(k).point1; lines(k).point2];
  plot(xy(:,1),xy(:,2),'LineWidth',2,'Color','green');
plot(xy(1,1),xy(1,2),'x','LineWidth',2,'Color','yellow');
plot(xy(2,1),xy(2,2),'x','LineWidth',2,'Color','red');

% Determine the endpoints of the longest line segment
len = norm(lines(k).point1 - lines(k).point2);
if ( len>max_len)
max_len = len;
xy_long = xy;
end
end


%% (contor + improved local binary pattern)
%Calculate Gradient Magnitude and Direction
Calc_G_M=im2bw(Input_Image);
[MG1, MG2] = imgradientxy(Calc_G_M);
[MG1_MAG, MG2_DIR] = imgradient(MG1, MG2);

Fig. (10), imshow(MG1_MAG); title('Output result for Gradient magni-
tude')
```

```
Fig. (11), imshow(MG2_DIR); title('Output result for Gradient direction')
Fig. (12), imshow(MG1); title('Output result for Directional gradient: X ax-
is')
Fig. (13), imshow(MG2); title('Output result for Directional gradient: Y ax-
is')
%% Design of Gabor Filters
%%%
INP_R_SZ = size(Input_Image);
NF_RWS = INP_R_SZ(1);
NF_CLMNS = INP_R_SZ(2);

W_LNGTH_MIN_VL = 4/sqrt(2);
W_LNGTH_MAX_VL = hypot(NF_RWS,NF_CLMNS);
RTO_VL = floor(log2(W_LNGTH_MAX_VL/W_LNGTH_MIN_VL));
Ipt_WLTH = 2.^(0:(RTO_VL-2)) * W_LNGTH_MIN_VL;

Input_t_VLU = 45;
Inpt_ORI_VL = 0:Input_t_VLU:(180-Input_t_VLU);

Input_FLTR_VL = gabor(Ipt_WLTH,Inpt_ORI_VL); % Filter
FLT_MAG_TD_VL = imgaborfilt(Input_To_Gray,Input_FLTR_VL);

% Post-process the Gabor Magnitude Images into Gabor Features
for i = 1:length(Input_FLTR_VL)
   SIG_VL = 0.5*Input_FLTR_VL(i).Wavelength;
   K = 3;
   FLT_MAG_TD_VL(:,:,i) = imgaussfilt(FLT_MAG_TD_VL(:,:,i),K*SIG_VL); %
noise reduction using filter approach
end

SC1 = 1:NF_CLMNS;
SR1 = 1:NF_RWS;
[SC1,SR1] = meshgrid(SC1,SR1); % Cartesian grid in 2-D/3-D space
ELT_FTR_set = cat(3,FLT_MAG_TD_VL,SC1);
```

```
ELT_FTR_set = cat(3,ELT_FTR_set,SR1);

% numPoints = N_RWS*N_CLMNS;
SC1 = reshape(ELT_FTR_set,NF_RWS*NF_CLMNS,[]); % Resize the image
based on the row and coloumn

SC1 = bsxfun(@minus, SC1, mean(SC1));
SC1 = bsxfun(@rdivide,SC1,std(SC1));

Inp_Co_Ef_VL = pca(SC1);
Extr_FTRE = reshape(SC1*Inp_Co_Ef_VL(:,1),NF_RWS,NF_CLMNS);
Fig. (14),imagesc(Extr_FTRE);

% Classify Gabor Texture Features

Input_Text_Analysis = kmeans(SC1,2,'Replicates',5);
Input_Text_Analysis = reshape(Input_Text_Analysis,[NF_RWS
NF_CLMNS]);
Fig. (15), imagesc(label2rgb(Input_Text_Analysis));

Inp_SEG_RSTS_1 = zeros(size(Input_Image),'like',Input_Image);
Inp_SEG_RSTS_2 = zeros(size(Input_Image),'like',Input_Image);
TXT_AN_BW_VL = Input_Text_Analysis == 2;
TXT_AN_BW_VL = repmat(TXT_AN_BW_VL,[1 1 3]);
Inp_SEG_RSTS_1(TXT_AN_BW_VL) = Input_Image(TXT_AN_BW_VL);
Inp_SEG_RSTS_2(~TXT_AN_BW_VL) = Input_Image(~TXT_AN_BW_VL);
Fig. (16), imshow(Inp_SEG_RSTS_1);
Fig. (17), imshowpair(Inp_SEG_RSTS_1,Inp_SEG_RSTS_2,'montage'); %
Display the segmented results

%% Texture feature extraction

TX1 = TXT_AN_BW_VL;
TX2 = 0.01;
```

```
TX3 = 0.03;
TX4 =  50;

%%

[Saliency_Map, Feature_Maps, ICA_Maps, Input_Image] = ...
          PGM_FL_3(Input_Image, []);

Fig. (18), imagesc(Input_Image);title('Preprocessed Image');
%          axis off;
Fig. (19), imagesc(Saliency_Map);title('Saliency Map');
Fig. (20), imagesc(mean(Feature_Maps, 3)); title('Mean Feature Map');

%% Feature extraction

corners = detectHarrisFeatures(Calc_G_M);
 [features, valid_corners] = extractFeatures(Calc_G_M, corners);

Fig. (21); imshow(Calc_G_M); hold on
plot(valid_corners);


%% CHARACTER EXTRACTION %%%
%% Convert to binary image
threshold = graythresh(G);
imagen =~im2bw(G,threshold);
image=im2bw(G,threshold); %%% binary image%
%% Remove all object containing fewer than 30 pixels
imagen = bwareaopen(imagen,30);
pause(1)
%% binary image
Fig. (22),imshow(~imagen);
title('binary image')
%% Label connected components
```

```
[L Ne]=bwlabel(imagen);

%% Measure properties of image regions
propied=regionprops(L,'BoundingBox');
holdon

for n=1:size(propied,1)
rectangle('Position',propied(n).BoundingBox,'EdgeColor','g','LineWidth',2)
end
holdoff
pause (1)
%% Objects extraction
Fig.,
for n=1:Ne
   [r,c] = find(L==n);
   n1=imagen(min(r):max(r),min(c):max(c));
imshow(~n1);
pause(0.5)
end

%% character extraction %%
inv=imcomplement(~imagen);
Fig. (23),imshow(inv);
[ro,co]=size(inv);
textal=inv(1:ro,61:(co-60));
Fig. (24) ,imshow(textal);
title('characters image');

se=strel('disk',3); % Structural element (disk of radius 1) for morphological
processing.
gi=imdilate(textal,se);
Fig. (25),imshow(gi);
title('characters');
```

```
%% FEATURE EXTRACTION %%%
image=bwmorph(image,'skel',inf); %% skeletonizing the characters in the
image%
image=PGM_FL_5(image); %%% selecting the boundary features%
stats=regionprops(bwlabel(image),'all'); %%% measurement of region
properties%
skel_size=numel (image);

area=stats.Area;
majoraxislength=stats.MajorAxisLength;
minoraxislength=stats.MinorAxisLength;

function IN_1 = PGM_FL_1(A,IN_2,GM_VL)


if ~exist('A','var') || isempty(A)
error('Input image A is undefined or invalid');
end


if ~exist('w','var') || isempty(IN_2) || ...
numel(IN_2) ~= 1 || IN_2 < 1
  IN_2 = 5;
end
IN_2 = ceil(IN_2);

if ~exist('sigma','var') || isempty(GM_VL) || ...
numel(GM_VL) ~= 2 || GM_VL(1) <= 0 || GM_VL(2) <= 0
  GM_VL = [3 0.1];
end


if size(A,3) == 1
```

```
  IN_1 = bfltGray(A,IN_2,GM_VL(1),GM_VL(2));
else
  IN_1 = bfltColor(A,IN_2,GM_VL(1),GM_VL(2));
end



function INP_3 = bfltGray(INP_4,INP_5,SGM_D_VL,SGM_R_VL)


[X_INP,Y_INP] = meshgrid(-INP_5:INP_5,-INP_5:INP_5);
G_VL = exp(-(X_INP.^2+Y_INP.^2)/(2*SGM_D_VL^2));


dim = size(INP_4);
INP_3 = zeros(dim);
for i = 1:dim(1)
for j = 1:dim(2)


iMin = max(i-INP_5,1);
iMax = min(i+INP_5,dim(1));
jMin = max(j-INP_5,1);
jMax = min(j+INP_5,dim(2));
      I = INP_4(iMin:iMax,jMin:jMax);


      H = exp(-(I-INP_4(i,j)).^2/(2*SGM_R_VL^2));


      F = H.*G_VL((iMin:iMax)-i+INP_5+1,(jMin:jMax)-j+INP_5+1);
      INP_3(i,j) = sum(F(:).*I(:))/sum(F(:));

end
```

```
waitbar(i/dim(1));
end


function INP_SG = PGM_FL_2(A)


MI_VL    = 5;
SGM_VL = [3 0.1];

MX_GRNT_VL     = 0.2;
SRP_VL_S  = [3 14];
QNT_LVLS     = 8;
MIN_ED_STRN = 0.3;

if ~exist('A','var') || isempty(A)
error('Input image A is undefined or invalid.');
end
if ~isfloat(A) || size(A,3) ~= 3 || ...
min(A(:)) < 0 || max(A(:)) > 1
error(['Input image A must be a double precision ',...
'matrix of size NxMx3 on the closed ',...
'interval [0,1].']);
end



INP_12 = PGM_FL_1(A,MI_VL,SGM_VL);



if exist('applycform','file')
  INP_12 = applycform(INP_12,makecform('srgb2lab'));
else
```

```
  INP_12 = colorspace('Lab<-RGB',INP_12);
end


[GX_VL,GY_VL] = gradient(INP_12(:,:,1)/100);
G_SG = sqrt(GX_VL.^2+GY_VL.^2);
G_SG(G_SG>MX_GRNT_VL) = MX_GRNT_VL;
G_SG = G_SG/MX_GRNT_VL;


E = G_SG; E(E<MIN_ED_STRN) = 0;


S = diff(SRP_VL_S)*G_SG+SRP_VL_S(1);


QB_VL = INP_12; dq = 100/(QNT_LVLS-1);
QB_VL(:,:,1) = (1/dq)*QB_VL(:,:,1);
QB_VL(:,:,1) = dq*round(QB_VL(:,:,1));
QB_VL(:,:,1) = QB_VL(:,:,1)+(dq/2)*tanh(S.*(INP_12(:,:,1)-QB_VL(:,:,1)));


if exist('applycform','file')
  INP_2 = applycform(QB_VL,makecform('lab2srgb'));
else
  INP_2 = colorspace('RGB<-Lab',QB_VL);
end

INP_SG = repmat(1-E,[1 1 3]).*INP_2;
```

Index

M. Rajalingam, *Text Segmentation and Recognition for Enhanced Image Spam Detection*, EAI/Springer Innovations in Communication and Computing,
https://doi.org/10.1007/978-3-030-53047-1

The manufacturer's authorised representative in the EU is Springer Nature Customer Service Centre GmbH, Europaplatz 3, 69115 Heidelberg, Germany. If you have any concerns regarding our products, please contact ProductSafety@springernature.com

Printed and bound by CPI Group (UK) Ltd, Croydon, CR0 4YY
30/09/2026
02225665-0001